VIDEO
Astronomy

OBSERVER'S GUIDES / SERIES EDITOR LEIF J. ROBINSON

REVISED EDITION

Steve Massey

Thomas A. Dobbins

Eric J. Douglass

SKY PUBLISHING CORPORATION

Published by Sky Publishing Corporation
49 Bay State Rd., Cambridge, MA 02138-1200, USA
SkyandTelescope.com

Library of Congress Cataloging-in-Publication Data

Massey, Steve.
Video Astronomy / Steve Massey, Thomas A. Dobbins, and Eric J. Douglass. — Rev. ed.
p. cm. — (Sky & telescope observer's guides)
Includes bibliographical references and index.
ISBN: 1-931559-09-0 (alk. paper)
1. Video astronomy. I. Dobbins, Thomas A., 1958 – II. Douglass, Eric J. III. Title.
IV. Series
QB126.M36 2004
522'.63 — dc22 2004049151

Printed and bound in Canada

Contents

Foreword

Although Vladimir Zworykin's 1923 invention of the iconoscope ushered in the modern TV era, it took a long time for video technology to catch on with astronomers, amateur or professional. Even in 1962, a decade after most U.S. homes had sprouted ugly antennas on their rooftops or "rabbit ears" in their living rooms, astronomer J. D. McGee wrote: "Television signal-generating devices are still of limited use to the working astronomer."

I can recall only one professional facility that was built, literally, around a video camera. That was in the 1960s when Corralitos Observatory in New Mexico could boast "brightness measurements good to 0.05 magnitude, and position determinations good to a second of arc."

A quarter century ago, at least for amateurs, a home video system was frightfully expensive and unmanageably bulky. And since ordinary operations were enough to challenge an electrical engineer, it's no wonder the technique didn't entice hobbyists! In a real sense, video astronomy has just arrived — thanks to the invention of the camcorder and affordable, powerful computers. The promise of video is now being quickly embraced and exploited by amateurs.

Video has always had alluring attributes. One was its much greater sensitivity than the photographic emulsions that were the standard detector when TV became widespread. Thus one could cut an exposure to gain time resolution. Another was the potential for achieving real-time variable contrast. And today, a video camera's digital output combined with off-the-shelf software make image sampling and manipulation easy to do.

I remember going to a meeting in the early '80s where I learned that someday "soon" amateurs would have robotic telescopes and take electronic images via charge-coupled devices. Even this technophobe could clearly see that a huge revolution was in the making, though I was nevertheless surprised at how quickly the solid-state, amateur-astronomy revolution happened.

The same kind of epiphany occurred again in the late '90s. I saw Ron Dantowitz demonstrate that video could yield one of the Holy Grails of astronomy: images as sharp as a telescope can produce. And David Dunham proved that a hand-held camcorder was a princely machine for precisely timing occultations — when stars suddenly disappear and reappear as the Moon or an asteroid eclipses them. These apexes in space and time resolution are the essence of video.

The first half of *Video Astronomy* delineates the equipment and processes needed to take successful images. The second half straightforwardly explains how to apply video techniques to a variety of bodies in our solar system and beyond. It's a complete package. Now it's time to get to work!

Leif J. Robinson

Leif J. Robinson
Editor Emeritus
Sky & Telescope

Preface

For Christmas when I was nine years old, my father presented me with my first telescope. It was a humble department store refractor, with a tiny, tabletop tripod, a fixed eyepiece, questionable optics, and came in a plain box with no misleadingly glorious pictures of planets or colorful depictions of deep-sky objects printed on the outside. I couldn't wait to use it. But, what to look at first? Like so many beginning observers, I didn't realize that the Moon might be an interesting subject — until my older brother pointed it out. Grateful for the suggestion, I perched the telescope atop a fruit crate and pointed it skyward on a Moonlit night.

That first glimpse of the Moon remains burned into my memory. I could see the magnificent splash of bright rays around Tycho, and other craters partially submerged in shadow near the terminator. The lunar landscape took on new reality for me as I peered across the gulf between Earth and Moon. I was seeing another world! For the first time I was looking at the night sky with curiosity and asking questions.

I started to keep meticulous notes and sketches from each night's observing session. About a week or two later, at an hour when all good boys should be sound asleep, I peered from the warmth of my bed through the window and noticed a bright star rising from behind some nearby trees. The allure was overwhelming. As the rest of the family slept, I tiptoed to the laundry room for a good vantage point, and to utilize the stable support for my telescope's tripod offered by the washing machine. As I focused my telescope on this bright temptation, a small disk appeared — the globe of Jupiter. At that moment I shared with Galileo Galilei the sense of wonder he must have felt the first time he pointed his telescope at Jupiter in 1610.

This was the beginning of a passionate interest that would persist through the years. No longer fascinated with cars, trucks, and plastic rifles, I dreamed of bigger telescopes, did more chores around the house to obtain the funds to buy them, and battled with my parents for later bedtime hours.

Astronomy and I parted ways for a few years while I played in a rock-and-roll band. Soon I was drawn back to the eyepiece and soothed my ringing ears under quiet night skies. This time I had a desire to capture faint, fuzzy objects in colorful photographs like those featured so prominently in popular astronomical magazines. As with most things, I taught myself through trial and error, eventually achieving some deep-sky pictures that made the observer's crick in my neck seem worthwhile. My obsession with the sky renewed, I resigned from a well-paid management position at a large information-technology firm to embark on a 16-month adventure to tiny, rural Coonabarabran in New South Wales, Australia. New South Wales is also the home of the Anglo-Australian Observatory at Siding Spring, where I was appointed to a support role for the telescopes' computers and information systems. It was a unique opportunity for any serious amateur astronomer. Under pristine dark skies and at a relaxed, unhurried pace, I could observe and photograph objects that I used to travel hundreds of kilometers on holidays to see.

While deep-sky photography at the prime focus of a telescope proved relatively easy, I never was able to obtain satisfying detailed photographs of the planets. Perhaps one print from each roll of film would be worth showing to others. Picking up these small, blurred, and often featureless results from the local film processor often brought a tinge of embarrassment. No doubt the badly drifted or jarred images led the photo shop staff to wonder, "Is this guy into photographing keyholes?"

After experimenting with a camcorder, I remembered that I had a tiny black-and-white, low-light video imaging circuit in my electronic junk box. I decided to wire it up and mount it to my telescope's focuser. The first results far exceeded my fondest dreams, producing sharp, detailed images of the Moon and planets. Nothing I had achieved with conventional photography had even come close. The experience rekindled the excitement I felt during my first childhood glimpses of the Moon and Jupiter.

Since the late 1990s video astronomy has literally exploded onto the amateur scene as a celestial-imaging alternative, taking its rightful place alongside conventional film photography and still-frame CCD imaging. I have no doubt that it will continue to yield marvelous results as camera performance improves and software capabilities evolve.

It is my fervent hope that this book will convey to the reader the simplicity of capturing wonderful celestial portraits with affordable, modern, and readily available video technology.

Clear skies!

Steve Massey

Acknowledgments

As the first edition of this book took shape, video technology advanced at such a rapid pace that it was bewildering, even frustrating, at times. Frequent changes and updates to the manuscript were required throughout the editorial process.

Today, amateur astrovideographers are now able to capture stunning portraits of deep-sky objects using new, highly sensitive video cameras, without the need for expensive image intensifiers. In the short time since *Video Astronomy* first went to press in 2000, we're pleased to provide this revised edition, which includes updated information, images, and references pertaining to the current state of amateur video astronomy. No doubt, advances in hardware and software technologies coupled with amateur ingenuity will continue to push the limits of video imaging.

This book would not have been possible without the wealth of information generously provided by various commercial firms, organizations, and, above all, like-minded individuals. The following are especially deserving of gratitude and recognition:

Ron Dantowitz of the Boston Museum of Science and Rob McNaught of Siding Spring Observatory shared their invaluable knowledge and experience.

The director and staff of Mount Stromlo and Siding Spring observatories allowed the use of a fine 24-inch telescope. Thanks are also due to Steve Quirk for his tireless assistance at the telescope.

Charles Genovese Jr., Michael Cross, Vince Ford and Steve Lee (Siding Spring Observatory), Jim Ferreira, David Frew, Gordon Garcia, Brian Govier, David Moore, Peter Neilson, Sandra Massey, Karen Dobbins, and Steve Wainwright all gave wise counsel and support, and in many cases allowed us to reproduce their images in this book.

Special thanks also go to John Cordiale of Adirondack Video Astronomy; to Jonathan Nally of

Sky & Space magazine; to Robert J. Richardson, who prepared the index; and to the team at Sky Publishing Corporation for all their hard work: Nina Barron, Dennis di Cicco, Rick Corson, Gregg Dinderman, Richard Tresch Fienberg, Sally MacGillivray, E. Talmadge Mentall, Lynn Sternbergh, Craig Michael Utter, Sean Walker, and especially to Carolyn Collins Petersen for her commitment to the project, valued advice, and for making it all come together. We'd also like to thank Edwin L. Aguirre for his courteous and proactive editorial assistance in updating this revised edition of *Video Astronomy.*

VIDEO ASTRONOMY

VIDEO
Astronomy

Clearer Skies Through Video Eyes 1

Has a friend or neighbor ever taken a peek at a beautiful image of Saturn through your telescope, only to step away and mutter a disappointed, “That’s nice”? Perhaps the focus wasn’t properly adjusted, or they failed to center their eye properly over the eyepiece, or they even bumped the telescope and knocked Saturn right out of the field of view. Maybe your grandmother always wanted to look through your telescope but wouldn’t brave the cold. This book offers solutions to problems like these. With affordable, user-friendly video technology you can record astonishingly clear, detailed views of the Moon and planets that will impress even the most seasoned observers, all at a fraction of the effort and expense of conventional film photography or still-frame charge-coupled device (CCD) imaging. And you can easily show your results to the rest of the world!

Starting with an ordinary camcorder, you’ll discover how easy it is to capture on videotape much of what you can see through your telescope. From this simple beginning, we’ll move on to more sophisticated video systems that will help you achieve even better results. We’ll discuss the fundamentals of video technology and survey the capabilities and performance of commercial cameras. You’ll even learn how to build an inexpensive, lightweight astronomical video camera on a rainy weekend. Finally, we’ll describe a series of observing projects, ranging from a few simple challenges for the novice to advanced video applications like fireball patrols, real-time deep-sky observing, and even high-resolution imaging of orbiting spacecraft.

This book is your passport to the exciting world of video astronomy. No longer will you have to settle for just describing what you saw at the eyepiece. You’ll be able to show everyone!

Short Exposures, Long Recordings

In recent years the advent of affordable CCD detectors has greatly diminished both the size and cost of video cameras while vastly improving their performance. A decade ago affordable video cameras were big and bulky, making them difficult to use with small telescopes. Today's cameras are compact, lightweight, easy to use, and remarkably sensitive to light. Furthermore, video offers the astronomer several terrific advantages over photography and still-frame CCD imaging. Let's begin by examining these advantages, then discuss the equipment you'll need to get started.

FIGURE 1.1
This image of a thunderstorm over Sydney, Australia, was captured using a camcorder. The upper and lower lightning strikes were captured on two consecutive 1/25-second frames, demonstrating the ease with which video can record such random, rapid events.

Most video cameras capture and display 25 or 30 images every second. Clicking and advancing a film camera 25 to 30 times each second to capture the scene in Figure 1.1 would surely qualify you for a place in the *Guinness Book of World Records*!

These short exposure times limit the amount of light collected by a video camera's detector. For this reason the brighter astronomical objects like the Sun, Moon, planets and many of their satellites, meteors, and brighter stars make good video targets. Faint nebulae, galaxies, and clusters are nearly impossible without the aid of an *image intensifier* — an instrument that amplifies the light coming from faint objects.

So why do video astronomy? What are the advantages of imaging suitable subjects with a video camera? Here are the principal advantages of using video to record your observations:

- **Exposure.** The short exposure times of video overcome many of the effects of the rippling and churning of our turbulent atmosphere. A continuous video recording of an observing session permanently captures remarkable details during fleeting moments of good *seeing* — when the atmosphere stabilizes enough to allow a steady view of an object.
- **Expense.** As a rule, video cameras are less costly and easier to operate than still-frame CCD cameras.

- **Telescope versatility.** Video systems can be used with a wide range of telescopes, including the popular Dobsonian reflectors that are not equipped with motor drives to compensate for Earth's rotation.
- **Viewing versatility.** Video is a dynamic medium that allows you to make dramatic recordings of events like eclipses, occultations, shadow transits of Jupiter's moons, and meteor showers.
- **Less eyestrain.** Viewing images on a video monitor reduces eyestrain and often increases visual acuity compared to peering through the eyepiece of a telescope. Many observers are able to discern more detail in images of the Moon and planets when they are displayed on a video monitor than by direct visual observation.
- **Convenience.** You can watch a video monitor in seated comfort. Gone is the need to perch atop a stepladder or bend over double to observe through the eyepieces of many backyard telescopes. To escape frigid temperatures or biting mosquitoes, you can send the output of a video camera to a monitor located indoors.
- **Group viewing.** Video is terrific for schools, museums, and planetariums. Interesting features can be simply pointed out on the monitor screen, without resorting to detailed verbal directions and instructions on how to focus the telescope.
- **Instant gratification.** With video you see results on the monitor in real time. There is no waiting for film to be processed to find out if you got the right exposure or tripped the shutter at the right instant.
- **Ease of image manipulation.** With a video processor or a computer and image-processing software, you can achieve some sophisticated techniques in an electronic darkroom without the messy chemicals!

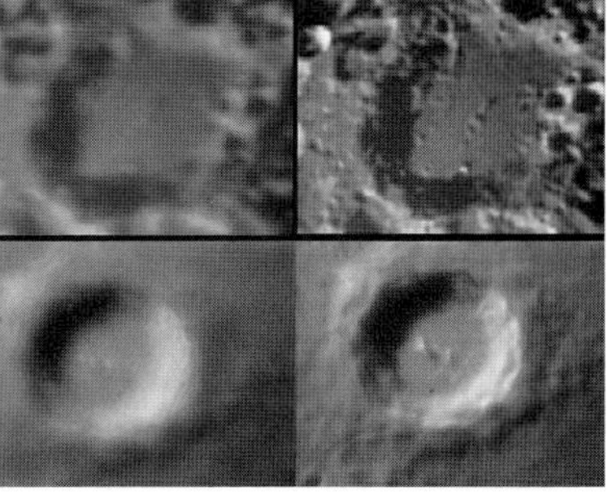

FIGURE 1.2
Atmospheric turbulence blurred three of these video frames of the imposing lunar craters Maginus (top) and Copernicus (bottom).

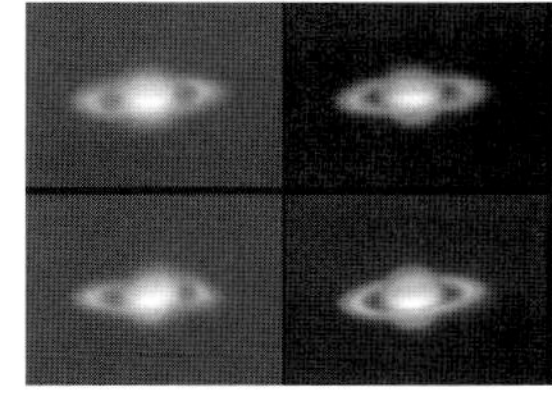

FIGURE 1.3
The effects of rapidly changing seeing are captured in this sequence of Saturn images sampled at 1/8-second intervals. Note the various distortions of the planet's appearance.

Die-hard purists may maintain that viewing a videotape is not experiencing the real thing. That may be true, but it's certainly the next best thing! A static image can never rival the emotional impact of a real-time moving video picture. A videotape of the Moon isn't just a permanent record of what you observed. It

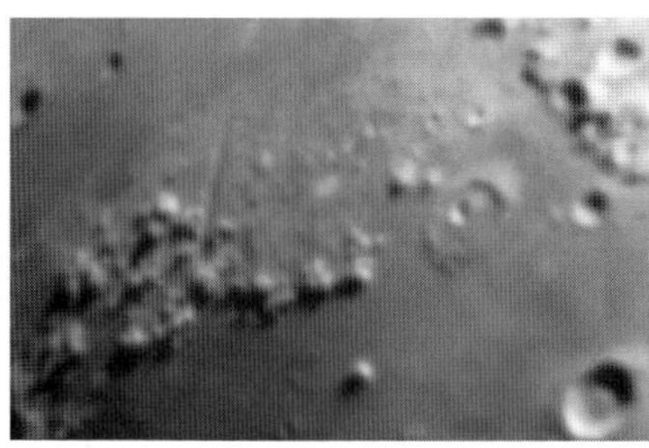

FIGURE 1.4
This conventional photograph of the lunar Alps was taken by an experienced astrophotographer under rapidly changing seeing conditions. Seconds before the shutter was snapped, the image had been far more sharply defined.

provides a unique, ineffable sense of reality that allows you to *relive* the night's observing. As atmospheric waves rippled across the stark beauty of the lunar landscape, you panned the telescope to survey every visible feature. You watched as shadow spires cast by towering mountains advanced or retreated across dark plains of frozen lava. The summits of peaks and the rims of craters flashed into view as they caught the first rays of the morning Sun. All these scenes, and more, can be captured with stunning and dynamic clarity rarely conveyed by a static image on a printed page.

What Equipment Do I Need?

To get your feet wet you only need a telescope, a video camera, a videocassette recorder (VCR), and a TV set or monitor. From this humble beginning, you can take the plunge and later acquire more sophisticated and specialized equipment. If you plan to use a system based around off-the-shelf components, you can adopt any of the following configurations:

- **A telescope and a consumer camcorder.** This simple combination can be mechanically awkward, but it does contain the minimum number of components. Although camcorders don't offer the levels of light sensitivity and resolution available in closed-circuit video cameras, many are capable of producing very pleasing results.
- **A telescope and a closed-circuit video camera.** This setup uses the integral recorder and viewfinder of a camcorder. It is an ideal arrangement for use at remote observing sites because it is extremely compact and powered by rechargeable batteries.
- **A telescope and a video camera in conjunction with a stand-alone VCR and monitor.** The advantage of this setup is that it is assembled from components that most observers already own.
- **A telescope and a video camera combined with a video capture device and a personal computer.**

After you've made a recording you can refine the quality of your images with a video signal processor, or tweak them on a computer using image-processing software. Hard-copy images can be made with a dedicated video printer or, if you've transferred the images to a computer, using a desktop printer. It's even possible to achieve excellent results by simply photographing the screen of a video monitor with a conventional 35-millimeter (mm) camera. The advantages and disadvantages of these various equipment configurations and techniques will be discussed in greater detail in following chapters.

Telescopes

Telescopes come in all sorts of shapes and sizes, and readers of this book are probably familiar with their operation. Most telescopes are capable of giving excellent results with video, though some are better suited than others. As described below, each design can pose unique challenges when the attempt is made to combine it with video equipment.

Refractors. Fine refractors deliver remarkably crisp, high-contrast images. However, they are very expensive per inch of aperture, and all suffer to some degree from an optical defect called *chromatic aberration* — the failure to bring light of all wavelengths to a common focus. The effect of this aberration can be glaringly obvious to most black-and-white video cameras, which can detect wavelengths of light out to about 900 nanometers (nm) or more in the near-infrared region of the spectrum. In most observing situations a Wratten #11 yellow-

FIGURE 1.5
With the price-to-performance ratio of personal computers continuing to plummet, closed-circuit video systems linking an automated telescope to a computer are being adopted by an ever-increasing number of astrovideographers.

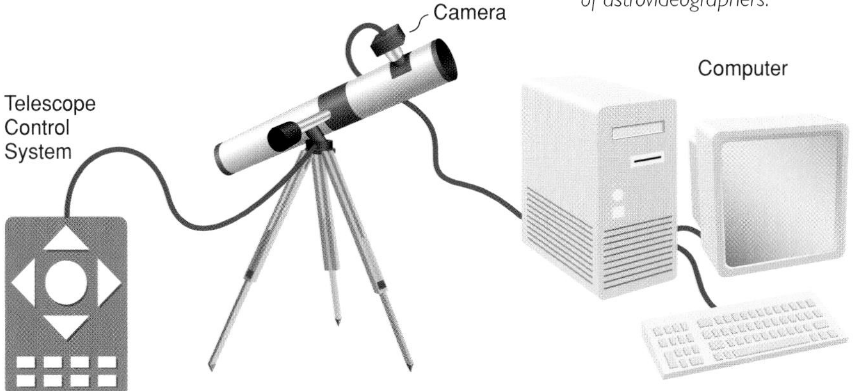

green or a Wratten #12 or #15 yellow filter attached to the camera will block the defocused wavelengths and sharpen the image.

FIGURE 1.6

A compact closed-circuit video camera can be used with even the most basic beginner's telescope like this humble 60-millimeter (mm) department store refractor.

Chromatic aberration is far less troublesome with apochromatic refractors, which have highly color-corrected objective lenses made with calcium fluorite or so-called *abnormal dispersion* glasses. However, the designers of these lenses are often primarily concerned with optimizing performance in the visible region of the spectrum from 400 to 700 nm. Consequently, even some apochromats produce the best video images only when a filter is used to restrict the wavelengths reaching the camera.

Newtonian Reflectors. The economical Newtonian reflector can rival the performance of any optical design. However, the location of the focuser at the side of the tube near its skyward end can be unduly awkward if a bulky camcorder is attached to an equatorially mounted Newtonian. Cantilevered at the end of such a long moment arm, its mass will require careful counterbalancing not only with respect to the declination and right ascension axes but also radially with respect to the tube.

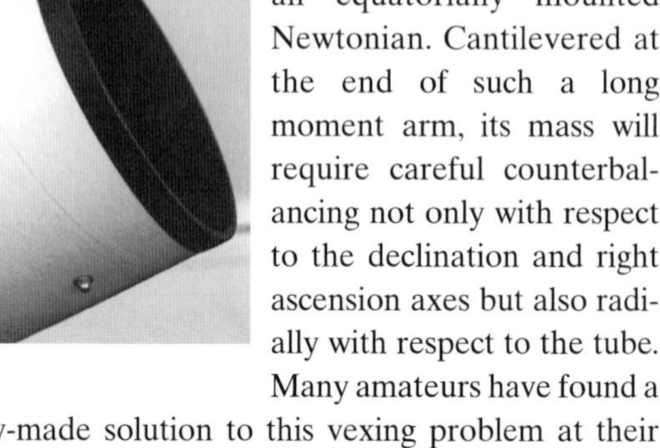

FIGURE 1.7

A video camera supported by the rack-and-pinion focuser of an equatorially mounted Newtonian reflector.

Many amateurs have found a ready-made solution to this vexing problem at their local sporting goods store in the form of jogger's ankle weights — metal shot sewn into handy, flexible bags that range in weight from ½ pound to 5 pounds. They can easily be attached to various locations on both the tube and mounting using strips of Velcro.

In recent years Newtonians riding on elegantly simple, low-cost altazimuth mountings called

Dobsonians have become very popular, allowing amateurs to acquire powerful instruments that would be prohibitively expensive if they were carried on conventional equatorial mountings. Unfortunately, Dobsonians can't follow the apparent motion of celestial objects imparted by Earth's rotation the way telescopes on equatorial mountings can. Dobsonians have long been considered the province of visual observers who must manually nudge the tube to keep their target in the field of view.

Still, if you own a Dobsonian or other altazimuth type of telescope, video imaging is definitely a tool you can use. Despite the drift of objects through the field, the fast frame rates of video will still produce glorious images. At low magnifications, the movement of celestial objects is less obvious, and an occasional nudge to recenter the image is all that's required. The limiting factor here is that as you increase the magnification, the field of view becomes smaller, affording ever briefer glimpses of your target. It also necessitates more frequent manual repositioning of the telescope. You'll soon find this tiresome and frustrating.

Dobsonian owners who yearn to do video work at high power can use affordable motor drive kits. These microcomputer-controlled systems can be retrofitted to your mount in a couple of hours using only hand tools. Another option is to buy or build a compact Poncet platform that will permit these scopes to follow objects for a couple of hours at a time. Both provide fully automatic tracking by means of electric motors.

Your video observing sessions will be far more pleasant and productive if your mounting is equipped with dual-axis motor drives and a hand-held, push-button controller. These features are all but essential if you intend to work at high magnifications with a video camera for any extended period of time. You'll be able to peruse the lunar surface while effortlessly panning around in any direction. Sitting back in your lounge chair and watching the lunar crater Copernicus fill three-quarters of your TV screen as you cruise by like the pilot of your own personal spacecraft is an experience not soon forgotten!

FIGURE 1.8
Left: *A video camera adapted for use with the Meade ETX 125-mm Maksutov-Cassegrain.* Right: *Observers with Dobsonian telescopes can record video using a tripod-mounted video camera aimed through the eyepiece of the telescope.*

Catadioptrics. The unsurpassed portability of the versatile Schmidt-Cassegrain and Maksutov-Cassegrain mirror-lens designs has made them extremely popular. However, their compactness comes with a price. Because these instruments employ primary mirrors with very fast focal ratios (typically f/2 to f/3), they are exceedingly sensitive to collimation errors, and their performance can suffer markedly if one of their collimation screws is off by even a small fraction of a turn. Poor collimation has undoubtedly contributed to the Schmidt-Cassegrain's reputation for producing images of indifferent quality. It's well worth the effort to collimate a catadioptric scope on a star at high power (40× to 60× per inch of aperture) until the diffraction rings appear perfectly round and concentric.

Most catadioptric telescopes focus by sliding their primary mirrors back and forth along a central light-baffle tube, driven by a threaded rod attached to a captive nut inside the focus knob. The resolution of this mechanism leaves much to be desired when trying to focus an image on the chip of a video camera. The smallest fraction of a turn of the focus knob spells the difference between a crisp image and a hopelessly blurred one. In addition, turning the focus knob on many of these telescopes produces a perceptible lateral displacement of the image in the field of view due to imperfect mechanical tolerances. This image shift is usually only a minor annoyance when the instrument is used visually, but it can be truly exasperating when imaging with a video camera at high power. If the highly magnified image of a planet leaps back and forth on the face of the sensor with every turn of the focus knob, achieving precise focus can be all but impossible. The solution is to invest in one of the elegant Crayford focusers that are offered by a number of firms to cure this very problem. The motorized versions of these focusers are highly recommended — you won't have to lay hands on the telescope and jiggle the image.

Astronomical Seeing

Regardless of the size and optical quality of your telescope, Earth's atmosphere is both the controlling and largely uncontrollable factor determining the clarity of the image the scope will produce on any given night, particularly at high magnifications. "The atmosphere," wrote the French astronomer André Couder, "is the worst part of the instrument."

The phenomenon that astronomers call *seeing* is caused by moving air cells in the atmosphere. They exist at all altitudes, from the surface to as high as 20 kilometers (km). These parcels of air have different temperatures and hence different indexes of refraction. Each cell acts like a lens, changing the focal position of the image in a telescope by bending incoming rays of light from a celestial object differently.

At most locations these seeing cells range in size from 10 to 20 centimeters (cm) in diameter, though research by atmospheric physicists and meteorologists has revealed that they can vary tremendously in size. When the aperture of a telescope is sufficiently large that it receives light that has passed through a mosaic of seeing cells, a blurred, "washed-out" image is the result. When the aperture is the same as the diameter of the seeing cells or smaller, the image is relatively unblurred, though focus changes as individual cells drift across the light path. The larger the aperture of the telescope, the smaller the probability that the air mass over it will be optically homogeneous at any given moment. This means that the moderate-aperture telescopes used by backyard astronomers make them very efficient instruments when it comes to revealing fine details on the Moon and planets. On many nights they can perform as well as professional telescopes many times their size.

Observing the Moon and planets has been compared to watching a movie with the projector out of focus except for brief, random intervals in which a few sharp frames occur. Seeing causes images to oscillate around a mean position in the field of view, a phenomenon called image excursion. The passing cells can also rapidly change the focus, producing a pulsating, fuzzy appearance. That's why visual observers have always been able to see and sketch much finer detail on the Moon and planets than any conventional photographer ever managed to capture on film. The human eye is a differentiating sensor, but photographic film integrates. As the eye studies the quivering image of a lunar crater or a planet, the brain rejects the poor images and concentrates on the brief moments when sharp, well-defined features are present. Photographic film, however, records both the good and the bad moments that occur during exposures that typically last up to 5 seconds. Almost invariably the result is a blurred picture.

Video mimics the capability of the human eye-brain combination to cope with the effects of atmospheric turbulence. Video cameras capture images every $\frac{1}{50}$ to $\frac{1}{60}$ second, and each image is output to tape or disk during the following $\frac{1}{50}$ or $\frac{1}{60}$ second, giving a rate of 25 to 30 images captured and displayed every second. The eye has a capture rate of 5 to 15 images per second, depending upon the brightness of the subject and variations from person to person. That's why video looks like a seamless continuum — 25 to 30 frames per second is above the threshold that is required for the phenomenon that psychologists call *flicker fusion* of a moving image to occur. Flicker fusion refers to the eye-brain combination that fills in portions of the picture that are absent for very short intervals. During this process noise is averaged out and frames of limited quality are smoothed over to create the perception of a sharp, vivid picture.

Both the visual observer and the photographer

must maintain an uninterrupted vigil so as not to miss those all-too-rare moments when the air steadies and details pop out. If a photographer fails to trip the shutter of the camera at just the right instant, the opportunity to capture a momentarily sharp view on film is lost forever. But with a video camera and a VCR it's possible to make a continuous recording during an entire observing session, permanently capturing those fleeting lucid moments like prehistoric bugs trapped in a piece of amber. At a remarkably affordable price a two-hour videocassette records hundreds of thousands of discrete images. At long last, here's a medium that combines the objectivity of photography with the dynamic sensitivity of the visual observer.

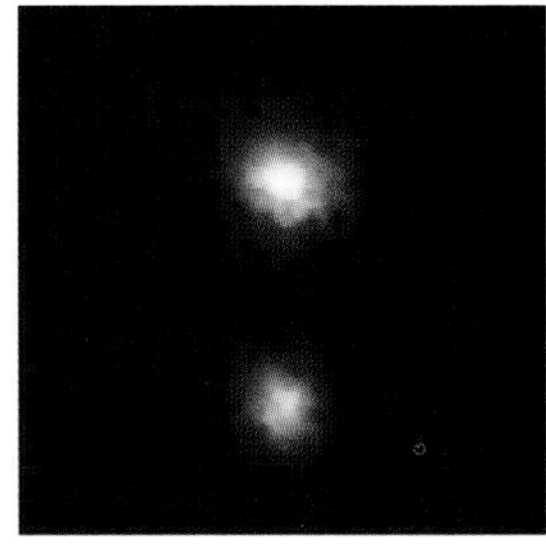

FIGURE 1.9
The rippling of the atmosphere is quite obvious in this true-color video image of the triple star system Alpha (α) Centauri. At an apparent elevation of only 25° above the horizon, the two brightest members look like a car's headlights seen through a curtain of running water. The 11th-magnitude red dwarf Proxima Centauri is the unseen member of this system.

Improving the Odds of Good Images

Many novices imagine that crisp, clear nights with stars twinkling like signal beacons are excellent for astronomical work. Unfortunately, on really transparent nights the air mass overhead is usually much colder than near the ground. This results in local convection currents and poor seeing conditions that persist until the air and ground temperatures come into equilibrium with each other. The resulting turbulence is what makes stars twinkle so much. Even deep-sky photography can be hampered because turbulence distends faint star images. This makes them appear fainter and require longer exposure times.

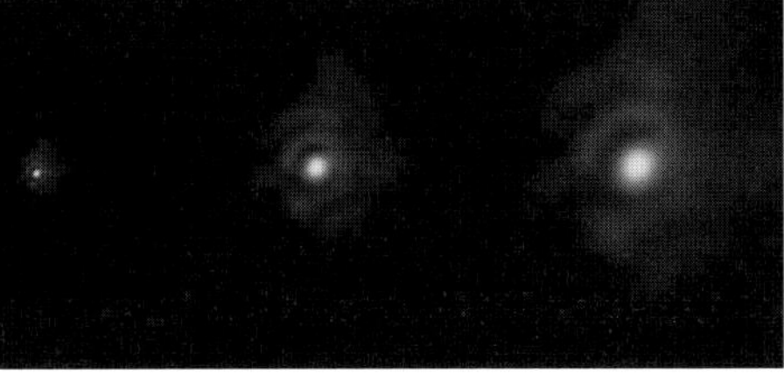

FIGURE 1.10
Even under rather turbulent atmospheric conditions, video is capable of capturing the diffraction rings that fleetingly appear around bright stars, as shown in this series of images of Vega through a 12-inch Schmidt-Cassegrain.

Good seeing is often accompanied by the presence of *haze,* a thin deck of high cirrus clouds, or a temperature inversion. These all retard ground cooling and the convection it produces. Sultry, humid nights when the telescope drips with condensation can be superb for getting really sharp images of the Moon and planets. Foggy nights can have the very best seeing (if transparency doesn't suffer too much) because the conditions that cause it to form require a very tranquil atmosphere.

Many observers have noticed that seeing often improves after midnight and is at its best just before dawn. The atmosphere can often be tranquil for a brief interval at evening twilight, when the variation of temperature with altitude (the *lapse rate*) is minimal. After nightfall heat is lost to space, and convection increases. As air masses begin to move, the seeing deteriorates. With the approach of dawn, thermal equilibrium is often re-established and the seeing improves again. Many planetary observers have enjoyed their finest views just after sunrise.

Virtually every geographical location has its share of tranquil nights, but some sites are far better than others. Good seeing tends to occur more frequently in tropical and subtropical regions due to the more uniform atmospheric pressures at lower latitudes and the absence of fast jet stream winds aloft. In temperate latitudes the best seeing conditions generally occur from the late spring through the autumn. Many winter nights are plagued by rippling and boiling images produced by turbulence in the jet stream.

Mountainous regions have traditionally been considered the best locations for good seeing because there is less overlying atmosphere to degrade the image. Unfortunately, winds blowing over peaks often create a locally turbulent air flow. Observing sites located on the slopes of mountains or in valleys are vulnerable to poor seeing caused by cold air descending from the surrounding higher elevations. Isolated plateaus tend to be excellent, and the seeing in a maritime region like coastal southern Florida is frequently superb, particularly when gentle onshore breezes provide a laminar air flow.

The effects of local topography on seeing should not be ignored. A grass-covered field makes an ideal observing site. Asphalt streets and parking lots, buildings, and rocky terrain devoid of vegetation all slowly radiate the heat that they accumulate during the day and should be avoided. The walls of an observatory should not be constructed of brick or cement block if these materials can be avoided. A simple, wooden roll-off roof structure usually has far better thermal properties than a more elegant and expensive dome.

On any given night the quality of the seeing usually varies with different objects, depending on their apparent elevation above the horizon. The main body of the Earth's atmosphere is about 16 km thick. The best seeing is generally found at the zenith, where the overlying blanket of air is thinnest. At an apparent elevation of 45°, the light from a celestial object passes through half again as much air, while the light of objects near the horizon must traverse 160 km or more of shifting, swirling air.

Aperture and the Odds

You can test for atmospheric turbulence at your observing site by throwing the image of a star or planet through your telescope out of focus and examining the distended image. Poor seeing will reveal a shimmering disk rapidly crisscrossed by flying shadows. Slowly undulating ripples indicate good seeing. On those rare nights when the seeing is truly superb, out-of-focus images are virtually motionless.

Writing in the December 1978 issue of the *Journal of the Optical Society of America,* physicist David Fried reported that the probability of experiencing a moment of diffraction-limited seeing decreases exponentially as the aperture of the telescope increases. On the bright side, however, the odds of capturing a sharp instant through telescopes of moderate aperture are surprisingly high. For a 12-inch telescope under typical seeing conditions, theory predicts that better than one image in 50 will be nearly perfect, so amateurs can achieve diffraction-limited results just by playing the numbers game. For a 24-inch telescope, however, the odds drop to perhaps 1 in 1,000. This may seem grim, but remember that a video camera can record a thousand discrete images in less than 20 seconds! As you can see, video offers powerful advantages over film or single-frame CCD cameras when it comes to high-resolution imaging.

Ron Dantowitz, an astronomy educator at the Boston Museum of Science, was curious as to whether video recordings were capable of capturing diffraction-limited images through a really big telescope, despite the diminished odds of success. Finding large, diffraction-limited optics proved challenging. Dantowitz's stunning images of Jupiter and Saturn that appear in this book were obtained with the

FIGURE 1.11
Jupiter in near-infrared light with an Astrovid 2000 video camera at the f/16.2 Cassegrain focus of the 60-inch Mount Wilson reflector. This impressively detailed image was assembled from 60 carefully selected video fields, or semiframes. Due to the appreciable diameter of Jupiter's disk, it was possible to select especially sharp sections from portions of the disk that appeared in various fields to create this breathtaking composite image.

famous 60-inch reflector at Mount Wilson Observatory overlooking Los Angeles. The optical quality of this 90-year-old telescope is excellent, and the air over Mount Wilson is often exquisitely steady. Dantowitz recalls that the images shivered and shuddered at the whims of the atmosphere for long intervals. However, when the seeing settled down the views were momentarily breathtaking. Jupiter sported swirls inside the Red Spot, and its belts and zones contained countless festoons and whorls. For brief instants Jupiter's Galilean moons were seen as beautiful, full little globes dappled with subtle markings. The subdivisions in Saturn's rings defied description, while Saturn's moon Titan appeared as a clean 0.7-arcsecond-diameter disk! Although the chances of getting lucky shots through a 60-inch aperture are fairly small, Dantowitz's superb images demonstrate that the benefits of video can be enjoyed by those fortunate enough to have access to a large telescope at a good site.

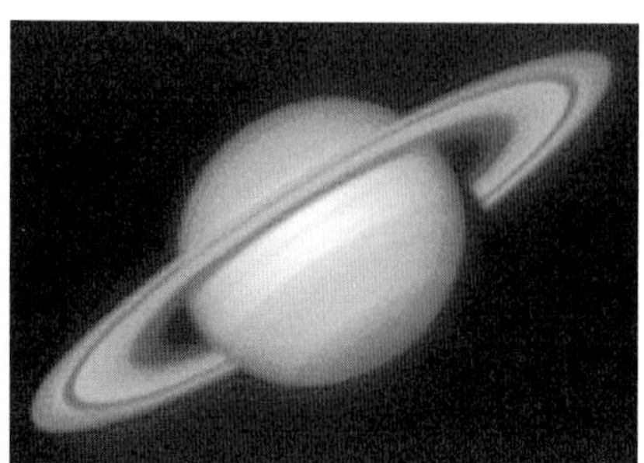

FIGURE 1.12
This image of Saturn is composed of 20 selected video fields that were recorded on August 27, 1997, using the same equipment that created the image of Jupiter in Figure 1.11.

How Video Cameras Work 2

Now that you've decided to do video astronomy, what's next? It will help to know how video cameras work and what performance specifications to consider when selecting one for astronomy. As you survey the marketplace, you'll find a bewildering array of equipment available, all descended from the vidicons of early TV broadcast equipment.

The Early Days of Video Technology

Television was invented in the 1920s but didn't become a popular medium for a quarter of a century. By the middle of the 1950s it was almost as common to find a TV set in the living rooms of American homes as a couch or chairs. During television's early years, all shows were broadcast live because recording technology lagged far behind the development of cameras, transmitters, and receivers. Indeed, some television shows were produced several times for viewing in different time zones!

The first videotape recorder (VTR), introduced by the Ampex firm in the 1950s, was an expensive and bulky machine. The early tapes were 2 inches wide and had to be threaded manually onto big studio reels, an irksome, time-consuming task. During the 1960s Phillips introduced the first VCR that housed tape in a convenient, easy-to-handle cassette. This set the stage for the debut of the first affordable, consumer-grade VCR in 1974.

Video cameras have also undergone a technological revolution. The first television cameras were heavy, cumbersome devices that recorded only in black and white and suffered from poor picture quality. For many years the heart of a TV camera was a vacuum television tube. This was followed during the 1950s by the far more light-sensitive vidicon and orthicon tubes. Because they were more compact than their predecessors, these advanced sensors made possible the first truly portable television cameras.

Television tubes contained a light-sensitive plate. When the light focused by the camera's lens fell on this plate, a photoelectric reaction caused the plate to release electrons. The darker areas of the image generated fewer electrons, while brighter areas generated larger numbers of electrons. The electrons then flowed to a second target plate, where they produced positive electrical charges. These charges were scanned by a gun that sent a beam of electrons directed by electromagnets to sweep the target in a succession of lines, read the local intensity of the charges, and register them as differences in the strength of an electrical current. This current was then amplified and transformed into a signal for transmission. The resolution of tube cameras left little to be desired, but they suffered from a phenomenon known as *lag.* If the scene or the camera was in rapid motion, the image smeared or appeared to leave a trail because the stored charge pattern on the photosensitive screen could not be entirely eliminated in a single scan.

CCDs

In 1969 physicists at Bell Laboratories invented the *charge-coupled device,* or CCD, a solid-state semiconductor. A CCD array consists of a thin wafer of very pure silicon that is subdivided into a grid of minute *picture elements,* or *pixels,* by a photochemical etching process. The pixels are very much like the tiles in a mosaic and act as light sensors. As photons strike a pixel they generate an electrical charge with a magnitude that is proportional to the intensity of the incident light. In a CCD video chip (Figure 2.1), each pixel acts as a tiny capacitor, accumulating a charge for a set interval called an *integration time.* Then the charge is spatially shifted to an adjacent storage register, which has the same opto-electrical properties as the active, light-sensing pixels. Storage registers are desensitized by placing a mask with opaque stripes over the surface of the array. Because the charges are transferred to vertical registers between columns of pixels, these CCD arrays are called *interline-transfer* arrays. Typically, only 30 to 50 percent of the array's surface is photosensitive, an inherent drawback of interline-transfer architecture.

As the pixels begin to collect the light of the next image, the packets of charges in the vertical registers are transferred column by column to a horizontal register that runs along the bottom of the array. This register is clocked by a driver circuit so that the voltage of each pixel is present in the output for only about one microsecond, during which time it is sensed by a transistor, amplified, and converted into an analog (voltage) video signal. The string of voltage pulses from each register is analogous to a single scan line of the electron gun in an old tube camera. Because the charges in an interline-transfer CCD array are shifted such a short distance to nearby registers and are read out in such minute fractions of a second, the resulting images are far less subject to lag than those provided by the old tube cameras.

FIGURE 2.1

The heart of a video camera: the CCD chip and the tiny window that protects it.

CCDs are remarkably efficient light detectors. About half of the photons striking the CCD array in a typical video camera generate an electrical potential that is converted into a signal. This performance is often referred to as a *detector quantum efficiency* of 50 percent. By comparison, for every 100 photons that strike a photographic emulsion, only a handful (generally two to four) interact with the emulsion's light-sensitive crystals of silver bromide to produce the grains of metallic silver that form a latent image. Even so-called "fast" films are 10 times less efficient than a CCD. The detector quantum efficiency of the human eye, by the way, is only about 5 percent.

However, CCDs are bedeviled by a problem known as *dark current.* During the integration time, random, thermally generated charges build up in each of the pixels, even in total darkness. At room temperature, this dark current can fully saturate a pixel after only a few seconds. Long before this stage is reached the recordable dynamic range (the number of brightness levels a chip can record) is reduced. The effects of dark current are minimized by the very short (1⁄50 second or less) integration times of CCD video cameras.

The chip in a CCD camera can be cooled with a thermoelectric device, known as a Peltier cooler. If an electric current is applied to two metal plates separated by a layer of semiconducting material, the combination acts as a heat extractor, transferring heat out of a CCD chip resting next to such a module. Each 5° to 6° C drop in the temperature of the chip decreases its thermal noise by a factor of 2. This drastically reduces dark current and permits integration times lasting minutes or even hours in order to record extremely faint sources of light.

Astronomers have been quick to employ this technique. With rare exceptions, cooled astronomical CCD cameras for recording faint objects have a frame-transfer chip architecture that cannot take the extremely short exposures that an interline-transfer video chip can. That's because the entire charge pattern of the array's pixels is shifted a far greater distance to a storage region that is identical in structure but completely covered with an opaque mask. It's like having two distinct arrays on the same chip — a light-sensing array and a memory array or matrix. The charges are then read out sequentially one row at a time instead of the odd- and even-interlaced method of interline-transfer video. Many of these cameras require mechanical shutters like those found in conventional film cameras in order to take very short exposures.

By now you may be wondering why frame-transfer CCD arrays are even made. In part, frame-transfer arrays are preferable because they lack the masked registers of interline arrays, so they collect more light and have higher spatial resolution per unit area. In addition, when cooled they have much lower dark-current noise and a much larger dynamic range.

Understanding Video Camera Performance

Black and white or color?

At first, it would seem preferable to have a color video camera. With it you could capture the ocher tints of the deserts of Mars and the delicate pastel hues of Jupiter's belts and zones. However, when we pause to carefully consider how color images are produced by a video camera, we find that its CCD turns out to be a modified black-and-white device! The modifications that permit colors to be distinguished inevitably sacrifice either light sensitivity or resolution. Because these are critical issues, we need to examine them closely.

The silicon of a CCD array generates an electrical charge that is proportional to the number of photons that strike it, but the array can't distinguish whether those photons are red, green, or blue — the three primary colors from which all other hues are derived. To permit the device to distinguish color, video engineers overlaid a CCD array with a striped, multicolored filter. The first row of pixels was covered by a red stripe, the second row by a green one, and the third row by a blue stripe. This alternating pattern was repeated across the entire face of the chip (Figure 2.2). To generate a color video signal, the light intensity information (i.e., the magnitude of the electrical charges) from the green columns was interpolated to assign green values to adjacent red and blue columns. This information was then used to construct a red-minus-green signal from the red-filtered columns and a blue-minus-green signal from the blue-filtered ones. A full-color image was finally derived by interpolating the relative intensities of these values.

The development of these single-chip color cameras paved the way for the multi-billion dollar consumer camcorder industry. Today's color cameras are the culmination of a long series of incremental improvements and often employ a more complicated mosaic filter of complementary colors that alternately transmit cyan, yellow, green, or magenta to the pixels rather than a simple series of alternating red, green, and blue stripes. The signals from four neighboring pixels are interpolated to generate full-color images. Although this ingenious method of obtaining color has been optimized to a remarkable degree, it does come at a price. Since multiple pixels must be interpolated to create a single point of color information, spatial resolution suffers. In stripe-filter cameras

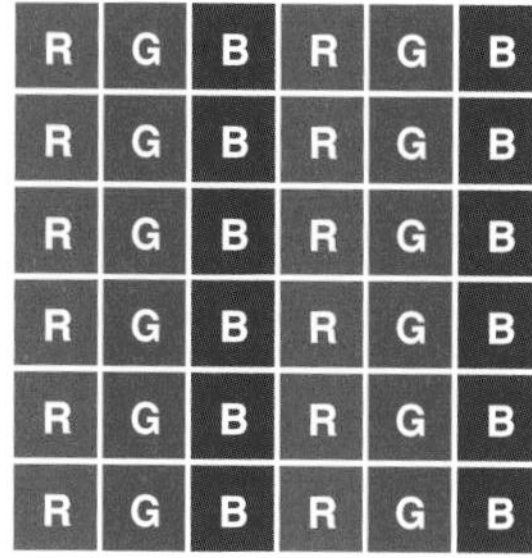

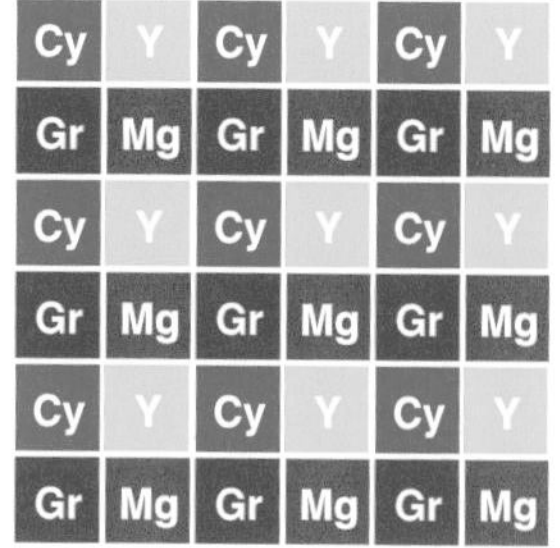

FIGURE 2.2

The configurations of the color filters on the CCD arrays of striped (left) and mosaic (right) single-chip color video cameras.

vertical resolution is preserved, but horizontal resolution is degraded. Mosaic-filter cameras provide improved horizontal resolution, but they still fall short of black-and-white cameras where each pixel corresponds to a single point of gray scale (an optical pattern of discrete steps or shades of gray between black and white) or brightness information.

Color decreases the sensitivity of the camera. Some light is lost in the color filters, and these cameras contain infrared-rejection filters that exclude the very part of the spectrum to which CCDs are the most sensitive. In addition the camera's signal-to-noise ratio (a measure of signal versus noise) tends to suffer because of the extensive electronic manipulation of the signal that color requires.

FIGURE 2.3

A three-chip color video camera uses dichroic coatings on the faces of prisms to separate the red, green, and blue components of incident white light and direct each beam to a separate CCD array.

The more sophisticated three-chip or RGB (red-green-blue) color cameras (Figure 2.3) employ a combination of prisms and filters to split the incoming light into three beams, which are then directed onto separate CCD arrays, each dedicated to a single primary color. The color information in the video output signal is kept on three separate channels, improving the signal-to-noise ratio. This method of creating color preserves spatial resolution, but at the expense of light sensitivity and much higher manufacturing costs, rendering these cameras unsuitable for many astronomical applications and most amateur budgets!

Black-and-white cameras are preferable for the

most demanding astronomical applications. Dollar for dollar they have superior resolution, contrast, light sensitivity, and signal-to-noise ratios than do color cameras. Bear in mind that the eye perceives spatial differences more clearly in gradients of gray anyway. In addition, applications that require video-to-computer interfaces may be compromised by color. Color images require considerably more processing time and more storage space in memory, yet often fail to yield significantly more information about the subject. If you absolutely require "one-shot" color images, then by all means purchase a color camera, but if you want to achieve the highest possible levels of resolution and sensitivity, stick with black and white.

Lux Ratings

Most of the objects in the sky are dim. To resolve fine details even on bright objects like the Moon and planets, you'll need to enlarge them to the point that they become dim! Video astronomy requires a camera with good light sensitivity so that you can employ the highest magnifications that the atmosphere and your telescope will allow.

The light sensitivity of a video camera is expressed in the *lux* rating system. The term is derived from the term *luminous flux* and defines the minimum intensity of light required to produce a usable picture. You might see this definition accompanied by an important clause that the chip (or faceplate) is illuminated by a very fast lens, typically with a focal ratio of about f/1.2. The definition is rather vague, and the same lux rating often means different things to different camera manufacturers. After all, just what constitutes a "usable picture?" Consumer magazine reviews of popular camcorder models have found little correlation between stated lux ratings and actual low-light performance. Despite this lack of uniformity and precision, however, lux ratings do provide a reasonable guide to light sensitivity. The first generation of camcorders typically had ratings of about 10 lux, meaning they needed a high light intensity to produce usable images. Many of today's affordable camcorders need much less light and thus are rated at 0.7 to 1.0 lux.

Black-and-white video cameras suitable for astronomy should have ratings of 0.1 lux or less, or about 1/1000 the level of illumination outdoors on a dull, overcast day! Equipment with a sensitivity of 0.05 lux or less is now readily available at affordable prices. Reasonably priced single-chip color cameras are generally rated at 0.5 to 3 lux.

You may encounter the light-sensitivity rating of a camera expressed in the rather antiquated unit of *footcandles*. One footcandle is equivalent to 10 lux, so a rating of 0.07 footcandles is equivalent to 0.7 lux.

A number of variables determine the faintest object that can be captured with a video camera attached to a telescope:

- the telescope's aperture, especially for point-source objects like stars
- the telescope's focal ratio, especially for extended objects like planets
- the light sensitivity (lux rating) of the camera
- the signal-to-noise ratio of the camera
- the quality of the video monitor and any recording device
- the presence of moonlight or light pollution.

As a benchmark, when an ordinary black-and-white surveillance camera rated at 0.05 lux at f/1.2 is mounted at the prime focus of a 10-inch f/4.5 Newtonian reflector, it will be capable of displaying stars down to magnitude 10.5 from a dark-sky site, even on a low-resolution monitor. Stars of this brightness can be seen visually through a humble 60-mm refractor.

Resolution

The more pixels in a CCD chip, the higher its resolution, theoretically. For example, if a chip of a given dimension is subdivided into 500,000 pixels, it will be capable of resolving much finer details than a chip of the same dimensions containing only 500 pixels. But there is a practical limit. If we continue to subdivide a CCD array into smaller and smaller pixels while holding the overall dimensions of the array constant, the light-gathering area of each pixel decreases and the array's sensitivity to light suffers. The individual pixels in most modern cameras measure about 10 microns across. Fortunately, video engineers have devised an ingenious way to reduce the tradeoff between more and smaller pixels and less light sensitivity by bonding a layer of microscopic lenses over the pixels to increase their effective light-gathering area. Thanks to these microlenses, the number of pixels in modern chips has increased even as lux ratings have fallen. Check for pixel numbers when you select a camera. Your camera's CCD array should have a minimum of 200,000 pixels.

However, resolution isn't merely a function of the number of pixels in a camera's CCD array. Mere pixel count can be very misleading if it is not interpreted correctly. Remember that the interline-transfer CCD arrays in video cameras do not contain perfect rows and columns of active pixels, and that there are masked registers where charges are transferred before they are read out. Even frame-transfer cameras with no dead spaces can yield imperfect center-to-center pixel resolution due to electronic noise and a phenomenon called *pixel jitter.* All of these factors contribute to the camera's spatial resolution (the amount of fine detail it can capture).

Bear in mind that the electrical charges generated when the pixels are exposed to light are converted into an analog (voltage) video signal. This video signal is received by the VCR, the monitor, and other components of the system. The resolution contained in the bandwidth of the video signal is determined in no small measure by a camera's internal electronic circuitry and is referred to as the camera's *TV line resolution.* The higher this number, the more capable the camera will be of revealing fine detail in the image.

But before you rush out and buy a camera with the highest TV line resolution rating that you can afford, pay attention to the resolution of the other components of your video system. For example, if you record with a VHS-format VCR, it will limit resolution to 240 TV lines, even if your camera puts out 400 lines. However, with a Super-VHS (S-VHS) or Hi-8 format recorder, you will be able to record 400 TV lines of resolution and take full advantage of your camera's resolution. For the very best results it is desirable that all of your equipment — camera, monitor, and recorder — be rated at 400 TV lines of resolution or more.

The superiority of CCDs over photographic film isn't simply the result of their higher quantum efficiency and increased sensitivity to light. A second and equally important factor contributes to the short exposure times with CCDs: to record fine details in images CCDs can work at shorter effective focal lengths and hence at faster focal ratios than conventional film cameras. With a typical CCD array's 10-micron pixels, only half the image scale (i.e., half the effective f/number) is needed to record the same level of detail that can be captured on a very fine-grained photographic emulsion. This corresponds to a fourfold decrease in exposure time.

FIGURE 2.4

The CCD arrays in video cameras are dwarfed by a frame of 35-mm format photographic film. Here is the Moon as it appears on (A) a slide taken at the prime focus of an 8-inch f/10 Schmidt-Cassegrain telescope. The rectangles illustrate the sensing areas of the following chip sizes: (B) ⅔-inch (8.8 mm × 6.6 mm), (C) ½-inch (6.4 mm × 4.8 mm), (D) ⅓-inch (4.8 mm × 3.6 mm), and (E) ¼-inch (3.2 mm × 2.4 mm).

CCD Array Formats

Most of the CCD arrays found in video cameras are tiny rectangles with a width-to-height or aspect ratio of 4 to 3. Confusingly, the format size of the chip expressed as a fraction of an inch is not equivalent to the sensor's active light-sensing area. Those actual physical dimensions for the most common chip formats are as follows:

- ¼-inch format: 3.2 mm × 2.4 mm (4.0-mm diagonal)
- ⅓-inch format: 4.8 mm × 3.6 mm (6.0-mm diagonal)
- ½-inch format: 6.4 mm × 4.8 mm (8.0-mm diagonal)
- ⅔-inch format: 8.8 mm × 6.6 mm (11.0-mm diagonal).

The active sensing area of the popular ½-inch format CCD array is almost 28 times smaller than a frame of 35-mm film (Figure 2.4)! At the prime focus of a telescope with a focal length of only 21.6 inches (550 mm), the disk of the Moon fills a ½-inch format chip. In fact the field of view displayed on the monitor using a ½-inch format video camera at the prime focus of any telescope closely approximates the field of view when the telescope is used visually with a high-power orthoscopic or Kellner eyepiece of about 7-mm focal length with a 45° apparent field of view.

FIGURE 2.5

Flip mirrors are very handy for locating and framing subjects in the narrow field of a video camera. They are available for standard 1.25-inch and 2-inch format accessories.

With these very narrow fields, merely locating an object can be quite an ordeal, even with a precisely aligned finderscope. Consequently, many astrovideographers augment their finderscopes with a medium-power telescope (often a 60-mm or 80-mm refractor) equipped with a cross hair reticle eyepiece that gives 25× to 50×. This combination allows quick acquisition of targets and avoids minutes of fruitless sweeping.

Others employ handy flip-mirror boxes to redirect the image from their eyepieces to their video cameras with the flick of a lever (Figure 2.5). These devices contain a movable flat mirror that can alternate between a "mirror down" position which reflects the incident beam of light 90° to an eyepiece for viewing and centering the target, and a "mirror up" position, which removes the mirror from the light path, allowing the incident beam to pass unimpeded through the box directly to the video camera.

Video Noise

One of the major problems in video astronomy is electronic noise. In addition to the noise inherent in a video camera's internal circuitry, random electrons spontaneously generated in its uncooled CCD array contribute a considerable amount of dark current noise. This can be further compounded by radio-frequency interference (RFI) picked up by imperfectly shielded cables as the camera's signal is transmitted to the VCR and monitor.

The noise characteristics of any electronic device like a video camera are denoted by its signal-to-noise ratio, which is expressed in decibels (dB). The decibel scale is not linear — noise of 20 decibels is 10 times as intense as noise of 10 decibels! You'll want a video camera with a signal-to-noise ratio of 45 decibels or more (often expressed as S/N > 45dB).

Removable Lenses

Many off-the-shelf video cameras (including most camcorders) are equipped with lenses that are not intended to be removed. These cameras give the astronomer no choice but to employ the afocal optical configuration (lining up the camera's lens with the eyepiece of the telescope, as illustrated in Figure 1.8) when coupling the camera and lens to a telescope-eyepiece combination. (Afocal imaging will be discussed in detail in Chapter 10.) A substantial loss of valuable light occurs due to the comparatively large number of optical elements involved. It can also be difficult to support the camera and keep it precisely centered and parallel with respect to the optical axis of the eyepiece in order to prevent distortion and unwanted ghost reflections. Finally, the lens cells of many of these video cameras (typically made of injection-molded plastic) do not have sufficiently robust mechanical construction to bear the weight of the rest of the camera when coupled to a telescope.

Fortunately, many video cameras (and virtually all

security and surveillance cameras) are supplied with removable C-mount lenses. They can be easily attached to telescopes using readily available T-to-C adapters. Just as the T-thread became the universal standard for 35-mm camera accessories, the C-thread is now widely employed for video lenses and accessories. (You may also encounter CS-mount video cameras or lenses. The diameter and pitch of the CS-mount thread —1-inch diameter and 32 threads per inch — is identical to the more commonplace C-mount, but the flange or back focal distance to the sensor is 12.5 mm rather than 17.5 mm for the C-mount.)

Removing the lens of a video camera exposes an antireflection-coated glass window that protects the fragile CCD chip. Even a tiny mote of dust on this window will look like a boulder on the screen, so keep it scrupulously clean using a cotton swab dampened with lens-cleaning solution. Take care to only dampen the swab; you don't want liquid to seep around the window onto the chip. A fine camel's hair brush from an art supply store can also be very useful for removing particles of dust.

You can avoid tedious, frequent cleaning of the window by obtaining a T-to-C adapter that is threaded to accept eyepiece filters. Use a clear Wratten #1A skylight filter to seal the camera-adapter assembly. Unlike the camera's protective window, the filter will be sufficiently far away from the chip that any dust specks that accumulate on it will be so defocused that they won't appear in the image.

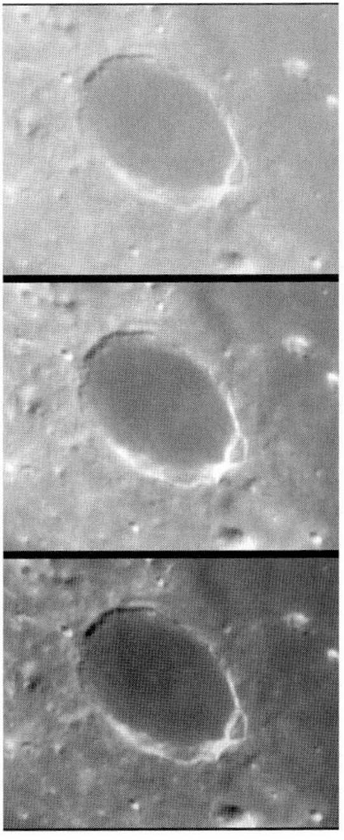

FIGURE 2.6

The effects of varying the shutter speed of a video camera are shown in these images of the lunar crater Plato. The top image is a 1/125-second exposure, the center is 1/250-second, and the bottom is 1/500-second long.

Shutter Speed

The electronic shutter in a video camera performs the same function as the mechanical shutter in a conventional film camera, controlling the amount of light to avoid overexposure or *blooming*. It does this by changing the integration time. Shutters in video cameras work in stepped increments, decreasing the amount of light collected by 50 percent for each increment. Most video shutters decrease exposure (integration) times in eight steps, from 1/50 or 1/60 second down to 1/10,000 second or less.

A camera that allows you to set exposure times

manually is preferable to one which automatically determines shutter speed. Cameras with automatic settings aimed at the Moon (Figure 2.6), for example, will grossly overexpose lighter areas, thus washing out features of interest. It's always better to have control over your exposure times.

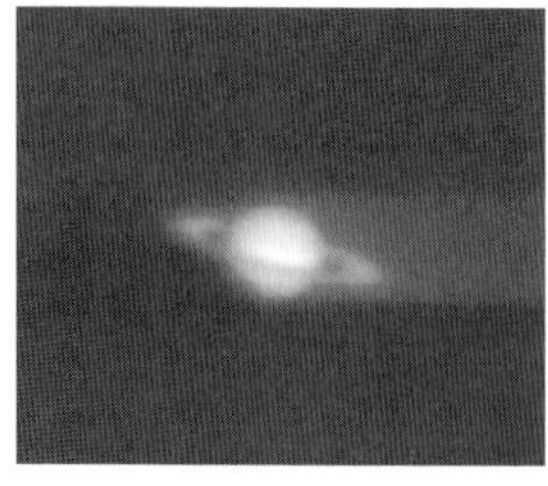

FIGURE 2.7

This image of Saturn was overexposed by an auto-gain, auto-shutter video camera. At right, the charges from saturated pixels have bled into adjacent pixels.

Automatic Gain Control

Automatic gain control (AGC) is an electronic feature that compensates for what the camera perceives to be low light levels by boosting the strength of the signal. If you're shooting a school play in a darkened auditorium, it's a nice way to get decent pictures of little Mary as she recites her lines on stage. It may also sound ideal for astronomy, but in fact the opposite is often true. Consider the case of planetary imaging cited previously. The camera will "see" the large expanse of dark sky background. The AGC will attempt to compensate by increasing the camera's sensitivity, compounding the overexposure problem. Furthermore, boosting the signal unavoidably increases noise, causing images to look grainy. Figure 2.7 shows the effect of using auto-gain and auto-shutter settings. Most video cameras have AGC. You just need a way to turn it off, so look for cameras with a manual AGC shutoff feature.

Selecting a Video Camera 3

Consumer Video Cameras

Camcorders come in a bewildering array of shapes, sizes, and degrees of sophistication. Sony, Canon, JVC, Panasonic, and the other leaders in camcorder technology offer models ranging from the most basic to units that rival the performance of professional TV studio equipment. Many include built-in editing features. Camcorders are designed to satisfy the requirements of home movie makers and are not optimized for the most demanding astronomical applications. Still, they have established a definite niche in astrovideography. Beautiful recordings of solar and lunar eclipses have been obtained with equipment no more elaborate than a camcorder mounted atop a sturdy tripod. Some readers may be so delighted with the results they obtain with a camcorder coupled to their telescope that they aren't even tempted to acquire more sophisticated video equipment.

FIGURE 3.1

A consumer camcorder can be a great starting point for capturing stunning footage of lunar landscapes and the planets.

The modern camcorder (Figure 3.1) consists of five principal components: zoom optics, a CCD array sensor, an integral videocassette recorder, a built-in monitor, and a rechargeable battery power source. Together they comprise a veritable movie studio in one incredibly small package.

Camcorders have an autofocus feature that, with rare exceptions, performs poorly under low-light conditions, frequently going in and out of focus. If you intend to use a camcorder, look for one with a manual override that disables the autofocus feature.

Despite remarkable advances in camera miniaturization, the recording tape transport mechanism, monitor, and battery of a camcorder all contribute substantial additional weight to an observing setup.

This may not prove troublesome for larger amateur instruments, but on most telescopes the weight of a camcorder adds undue stress to the focuser. Many amateur astronomers have fabricated brackets that attach to the tube of their telescopes to provide an adjustable camcorder support platform. Universal camcorder-to-telescope brackets like this are also offered for sale by a number of firms. Whether home-built or store-bought, they are a vast improvement over holding a camcorder up to the eyepiece by hand! Once the problem of coupling a camcorder to the telescope is solved, one additional challenge remains: careful counterbalancing to prevent slipping the drive's clutch or overtaxing the motor.

Some of the newer camcorders boast remarkable 220× digital zoom capabilities and "zero lux" light sensitivity. These impressive specifications sound ideal for astronomical applications. Just how does such a tiny camera manage to magnify the image a staggering 220×? You'll be disappointed to learn that these cameras are only capable of magnifying up to 22× by optical means. The secret to achieving even higher magnifications is an electronic algorithm called *digital image enlargement* that doesn't increase resolution. Long before a magnification of 220× is reached, the checkerboard pattern of pixels that make up the image becomes glaringly obvious.

The zero lux feature also proves disappointing astronomically. When it is engaged it disables the chip's infrared-cut circuit, making the camera sensitive to infrared light. The camera then emits an invisible beam of infrared light to illuminate nearby subjects. But that beam won't perceptibly illuminate the distant Moon any more than the flash of a film camera will!

Security and Surveillance Cameras

Manufacturers of video cameras for low-light security and surveillance applications were quick to capitalize on the excellent light sensitivity of CCDs. By avoiding the filters required for color imaging, they were able to produce black-and-white cameras with

FIGURE 3.2
Virtually all security and surveillance cameras feature interchangeable C-mount lenses, making them easy to couple to a telescope. Their excellent light sensitivity makes them ideal for recording many astronomical subjects. Some models offer manual exposure controls and override of automatic gain control (AGC).

superb light sensitivity that are well-suited to many astronomical applications.

There are a host of lightweight, compact security and surveillance cameras available that feature removable lenses and accept standard C-mount accessories (Figure 3.2). Most units feature die-cast or extruded aluminum housings equipped with an external power indicator, power switch, electronic shutter control, and composite or BNC video connectors (more on these later). Sensitivity varies from 0.01 to 0.1 lux in the black-and-white models and as low as 0.5 lux in the single-chip color models. These cameras are designed specifically for low-light imaging and are capable of producing outstanding results in astronomy. Their spectral response at both visible and infrared wavelengths make them ideal for most astronomical applications.

Here is a typical specifications list for monochrome security video cameras.

Image sensor:	⅓-inch Sony HAD ICX054AL
Image sensor size:	4.9 mm (H) × 3.7 mm (V) Pixel = 9.6μm (H) × 7.5μm (V)
CCD pixels:	ALL: 510 (H) × 492 (V) EIA 500 (H) × 582 (V) PAL
Horizontal frequency:	15.750 kHz (EIA) / 15.625 kHz (CCIR)
Vertical frequency:	60Hz (EIA) / 50Hz (CCIR)
Scanning system:	Interlaced
Minimum illumination:	0.04 LUX @ f/1.2
Resolution:	410 Lines

Signal-to-noise ratio:	>48 dB (AGC off)
Electronic shutter:	1/60 (1/50) to 1/100,000 sec.
Signal output:	Composite 1.0 Vp-p 75 ohm
Gamma settings:	0.25 , 0.45, 1.0
Operating temperature:	–20° C to 50° C
Operational humidity:	Within 80 percent MRH
Power required:	10 to 14 VDC @ 250mA (110 or 220/240 VAC plug pack)
Dimensions:	60 mm (W) × 70 mm (H) × 100 mm (L)

Purpose-Built Astronomical Cameras

While there are many camera options to choose from, purpose-built astronomical designs offer superior results quickly and conveniently. They are supplied with all the necessary controls, couplings, and cables required for trouble-free setup and out-of-the-box operation.

Adirondack Video Astronomy (AVA) of Hudson Falls, New York, is a leading supplier of astronomical video cameras. Their Astrovid line of cameras is widely employed by amateur and professional observers. The basic model, the Astrovid 1000, weighs in at around 350 grams (12.5 ounces) and employs a 1/3-inch-format, 0.02 lux image sensor made by Sony that produces 410 TV lines of resolution. Another feature of particular interest to occultation observers is a built-in microphone that can be used for timing purposes in conjunction with a reliable time reference like station WWV. The more sophisticated Astrovid 2000 series cameras use a 1/2-inch-format Sony CCD array that contains 768 horizontal rows and 494 vertical columns of active pixels and delivers 600 TV lines of resolution for outstanding picture clarity when it is used in conjunction with a high-resolution television monitor (Figure 3.3).

While most commercial surveillance cameras allow the user to adjust their gain (sensitivity), gamma (contrast), and shutter speed (exposure) settings, these switches and control knobs are almost invariably located on the camera's housing, a logical place indeed for the controls of a fixed surveillance camera. However, a good shielded cable can carry a video signal from a camera for distances up to 75 feet to a monitor or VCR. (Even longer distances are possible if a signal booster is added to the line). For an astro-videographer operating a camera remotely from indoors, control knobs located on the camera itself mean a lot of running back and forth to adjust the camera's settings! AVA designed the Astrovid systems with this in mind and supplied their cameras with a remote-control box to permit the user to remotely adjust the video camera's gain (sensitivity), gamma (contrast), and shutter speed. AVA offers accessory extension cables in lengths up to 50 feet. The latest Astrovid 2000 series camera features a convenient 4-inch liquid crystal display (LCD) monitor built into a larger hand-control box. Figure 3.4 shows an image taken with the Astrovid 2000.

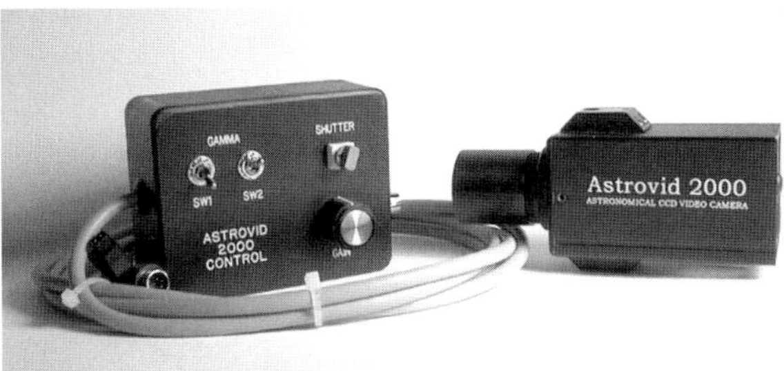

FIGURE 3.3
The popular Astrovid 2000 camera features remote and manual controls for shutter speed, gain, and gamma (contrast) adjustment.

While the black-and-white Astrovid cameras have emerged as the instruments of choice for serious astrovideographers, Adirondack has also developed a family of single-chip color cameras with sufficient light sensitivity to provide surprisingly good performance on the brightest subjects, even with telescopes of very modest aperture. The first camera in this series, known as the Planetcam, was introduced early in 2000. Smaller and lighter than most 1¼-inch-format eyepieces, it employs a ¼-inch-format CCD array with tiny 4.2 × 4.2 micron pixels. While the Planetcam lacks the features required for the most demanding applications — user-selectable shutter speeds and the ability to disable AGC — it is very affordably priced, easy to use, and provides pleasing but by no means state-of-the art results. The Astrovid Color Pro series cameras,

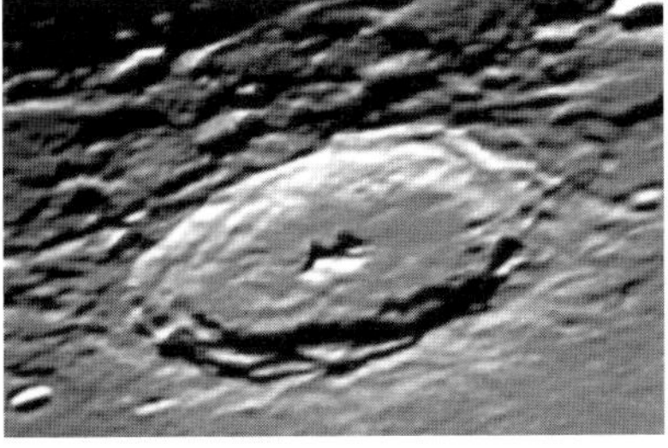

FIGURE 3.4
This image of the lunar crater Langrenus was made with an Astrovid 2000 camera using eyepiece projection with a 10-inch Newtonian reflector. It demonstrates the satisfying results that can be obtained with this camera and a typical backyard telescope.

based on designs originally intended for research microscopy, incorporate all of the features of their black-and-white counterparts but fetch three to five times the price.

Vixen Optical Industries, Japan's leading manufacturer of astronomical telescopes and accessories for amateurs, offers two purpose-built astronomical video cameras. The black-and-white model B05-3M (Figure 3.5), built around a 291,000-pixel $\frac{1}{3}$-inch-format CCD array, boasts an impressive light-sensitivity rating of 0.05 lux and provides 380 TV lines of resolution. Weighing in at a mere 200 grams (7 ounces), it offers manual gain and contrast controls, but both are located on the body of the camera itself. Vixen's C10-4M model is a color camera with a tiny $\frac{1}{4}$-inch-format CCD array and a light-sensitivity rating of 1.0 lux that provides 330 TV lines of resolution. Both cameras use a 12-volt DC power supply.

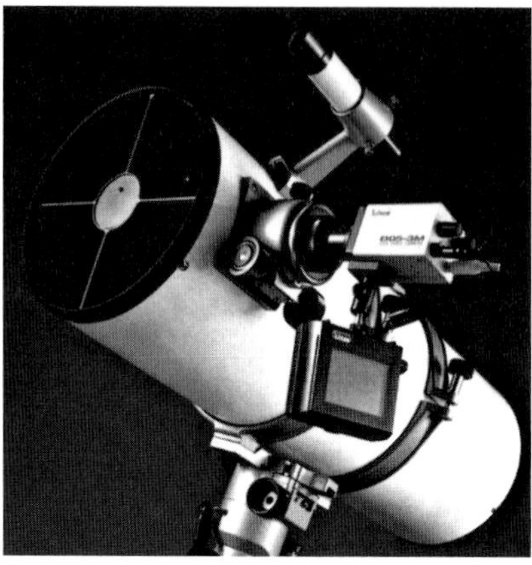

FIGURE 3.5
Vixen Optical Industries' model BO5-3M video camera, shown here with an accessory liquid crystal display (LCD) for convenience of viewing near the focus of an 8-inch f/4 Newtonian reflector.

The Digital Revolution

We're on the verge of a revolution! Digital video cameras are beginning to displace their analog counterparts in many industrial, scientific, and medical imaging applications. The first generation of digital consumer camcorders appeared on the market in 1995. It's only a matter of time before prices fall to the point that digital cameras largely supplant analog systems and forever change the way we think about video.

In a digital camera, the charges from a CCD array are converted to digital data inside the body of the camera prior to sending the signal to a monitor, recorder, or processor. This allows for the highest fidelity by reducing the chances that electromagnetic and radio-frequency interference (EMI/RFI) will contaminate the signal.

The voltage impulses of an analog video signal are waves that can be represented as the up and down movement of a line. They look like a succession of peaks and valleys, with gradations in amplitude (the heights of the peaks and the depths of the valleys) and frequency (the distances between the peaks and valleys). These variations correspond to white and black and all the levels of gray in between. Electrical interference can change the height of the peaks or the depths of the valleys, as well as introduce spurious peaks and valleys that weren't present in the original signal, resulting in a loss of fidelity.

Analog video recording is also subject to timing errors introduced by imperfect tolerances in the mechanical components involved. If you copy an analog videotape with an analog recorder, the duplicate (second generation) will have poorer picture quality than the original. Make a copy of a copy (third generation), and picture quality will be further degraded. Each succeeding generation removed from the original results in progressively greater loss of sharpness and color.

The electrical signals of a digital camera consist of just two values: on and off. This binary code (really ones and zeros) provides a robust signal that can be transmitted and copied repeatedly without distortion. But with an analog system an infinite number of values are possible, making it difficult, if not impossible, to distinguish valid values from erroneous ones.

Digital image data can also be stored in a memory buffer inside the camera. In addition, adjustments to brightness and contrast, and even image-processing routines, can be performed on the fly to improve the image data before it reaches a host processing system. By comparison, only very rudimentary image processing can be performed in the analog domain. Image data straight from the camera is ready for real-time image processing using custom electronic hardware or personal computers. While a digital data stream can be transmitted without degradation via the common RS-422 protocol, the manufacturers of digital video (DV) cameras are now embracing the IEEE-1394 standard, otherwise referred to as iLINK or Firewire, the name chosen by Apple (its developer). This new standard is a welcome improvement over the clunky Small Computer

System Interface (SCSI) technology and offers much faster data-transfer rates than the popular Universal Serial Bus (USB).

Progressive-scan digital cameras acquire an entire image with a single electronic shutter event in their CCD sensor. They output an entire frame of image data one line at a time with frame rates that range from 12 to 1,000 frames per second, depending on their horizontal and vertical pixel resolution. These cameras have a marked advantage over interline-transfer cameras in stop-action situations (like "freezing" astronomical turbulence), motion-analysis applications, and tracking fast-moving objects. In some rolling mills, for example, progressive-scan video cameras are used to inspect sheets of aluminum for defects while the product is moving at speeds of more than 200 feet per second!

Digital progressive-scan cameras are currently in widespread use for airborne imaging, security, automated inspection, machine vision, license-plate recognition, and traffic-control equipment. They have already eclipsed analog video systems in areas of speed, real-time image processing, and precision imaging.

However, the additional onboard processing electronics of digital cameras increase their size, power requirements, and the cost of the design and manufacturing when compared with analog cameras. In essence, though digital cameras are ideal for the most demanding scientific applications due to their convenient data format and image-processing capabilities, analog cameras remain desirable when small camera size and, above all, low cost are required. Using analog video cameras, backyard astronomers on hobbyist budgets can accomplish imaging feats that would have been regarded as nothing short of miraculous as little as a decade ago.

Desktop Computer Cameras

Commonly known as *video-conferencing cameras,* or *webcams*, miniature desktop digital video cameras have become a very popular, low-cost tool for astronomical imaging. Today, there are numerous brands and models available, each employing the sophistication and convenience of "plug-and-play" connectivity with desktop or laptop computers via a USB port. Aside from fast download speeds, another major benefit of USB-based webcams is that the electrical power needed for the cameras' internal electronics to function is derived from the computer, rather than the sometimes-less-convenient external power adapter needed for many standard security video cameras.

FIGURE 3.6
Webcams, such as the new Philips ToUcam Pro II (PCVC 840K), excel in lunar and planetary imaging. The ToUcam Pro II retails for around $140.

While many webcams have their own specific, proprietary features (some of which are useful while others are not), one name and model in particular seems to stand out from the rest. The Philips ToUcam Pro series has been sweeping the astronomical planetary imaging world by storm since it was introduced some two years ago. It's the camera known to produce spectacular pictures of the planets, such as Mars, Jupiter, and Saturn, many of which are often featured in *Sky & Telescope* magazine.

Weighing a mere 100 grams, the ToUcam Pro II model is enclosed in a small, compact plastic housing and features a 640-by-480-pixel CCD sensor. Some commercial dealers supply the camera with a convenient 1.25-inch eyepiece adapter for mounting it directly to the telescope's focuser (usually with a Barlow lens added for increased magnification).

FIGURE 3.7
Singaporean amateur Tan Wei Leong obtained this view of Io and its shadow (right of center) transiting Jupiter on February 11, 2003, with a 250-mm Takahashi Mewlon reflector, a 2× Barlow, and the now-discontinued Philips ToUcam Pro PCVC 740K webcam (see the book's back cover). It is a stack of 500 1/25-second exposures. Note the planet's Great Red Spot (upper left of center).

If you have good mechanical skills, you can, of course, make careful modifications to the camera yourself to fashion a suitable new housing or telescope attachment. This has been a common undertaking among amateur astronomers for years, particularly with the Connectix QuickCam PC webcam shown in the following pages. Bear in mind, however, that making any modifications to the camera will void its manufacturer's warranty, so proceed with caution.

The ToUcam Pro II offers full 24-bit color "still-image" capture, with interpolated resolutions of up to 1,280 by 960 pixels. Snapshots can even be activated by voice! Furthermore, this camera has a digital video resolution of 640 by 480 pixels and frame rates of up to 60 frames per second. While the minimum light sensitivity of the camera's sensor is rated at less than

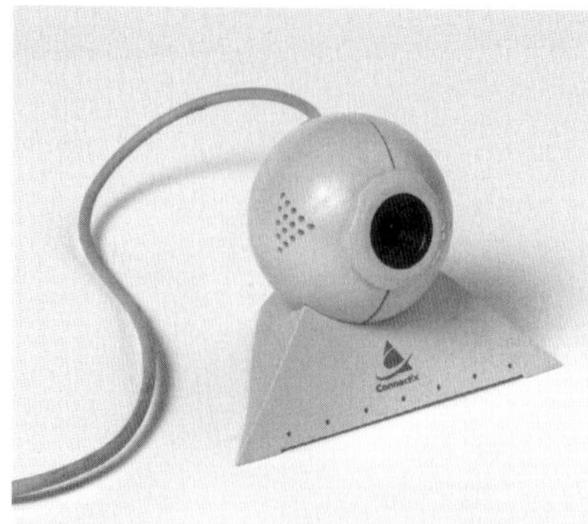

FIGURE 3.8
The QuickCam PC webcams can be easily modified for use with a telescope, providing an inexpensive tool for trying out electronic imaging.

1 lux, bright celestial subjects can be recorded beautifully. By stacking dozens of still images with programs such as *Registax*, the image's signal-to-noise ratio can be greatly improved to produce a smoother, more detailed picture. The resulting composite can then be processed and enhanced further with graphics-editing software such as *Photoshop* or *Paint Shop Pro*.

Many webcams come with USB cables between two and three meters in length. If your setup requires a longer cable, interconnecting extensions are available from major computer stores.

Nearly all webcam packages are bundled with software that allows you to vary the image's brightness, contrast, exposure times, gamma level, and color balance. Some even provide post-capture image-processing programs so you can further improve the results. You can also capture images directly from the camera into your favorite image-processing application via your computer's TWAIN driver. You can then save the images in a variety of formats, including BMP (.bmp), TIFF (.tif), JPEG (.jpg), and AVI (.avi).

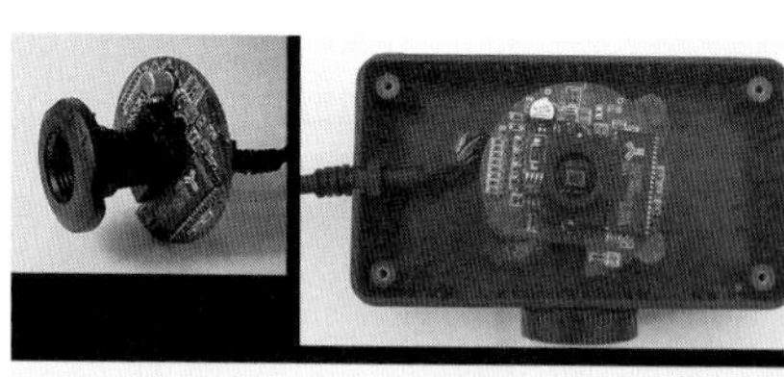

FIGURE 3.9
A modified QuickCam PC camera being installed in a project box enclosure.

An Exciting but Uncertain Future

Our first glimpse into the exciting future of video astronomy came in late 1999, when the Santa Barbara Instrument Group (SBIG) introduced the STV, a self-contained integrating digital video camera and autoguider. The STV features a thermoelectrically cooled frame-transfer CCD array (similar to those found in still-frame astronomical CCD cameras) for far-lower dark current and a larger dynamic range compared to uncooled video systems. As a result, STV users can select exposures up to a whopping 600 seconds long. This allows them to record faint objects far beyond the grasp of conventional video cameras, which are limited to maximum exposure times of $1/50$ to $1/60$ second.

The STV's black-and-white video output can be viewed on an external video monitor or on SBIG's optional 5-inch LCD (liquid-crystal display) monitor. The images can also be recorded on a VCR (with some loss of signal quality) or stored in the camera's onboard memory, where they can be downloaded later to a computer for viewing and/or processing.

The refresh rate for the video display depends on the length of the exposure selected. For exposures longer than one second, the display will not be updated until the current exposure is completed (the previous frame will continue to be displayed until it's replaced by the new image). For short exposures, the frames are displayed at a maximum rate of 16 frames

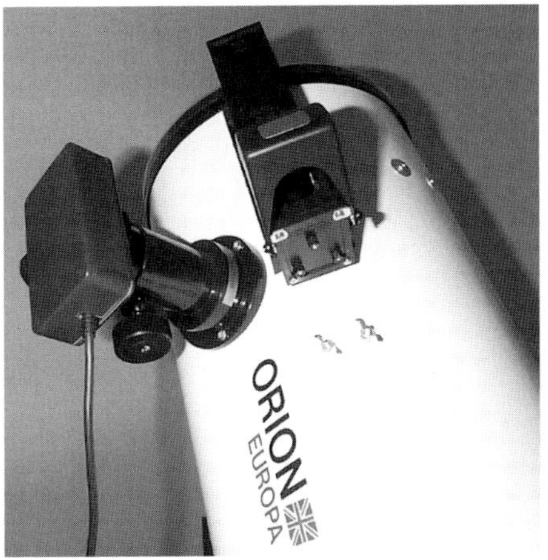

FIGURE 3.10
A rehoused QuickCam mounted and ready for use.

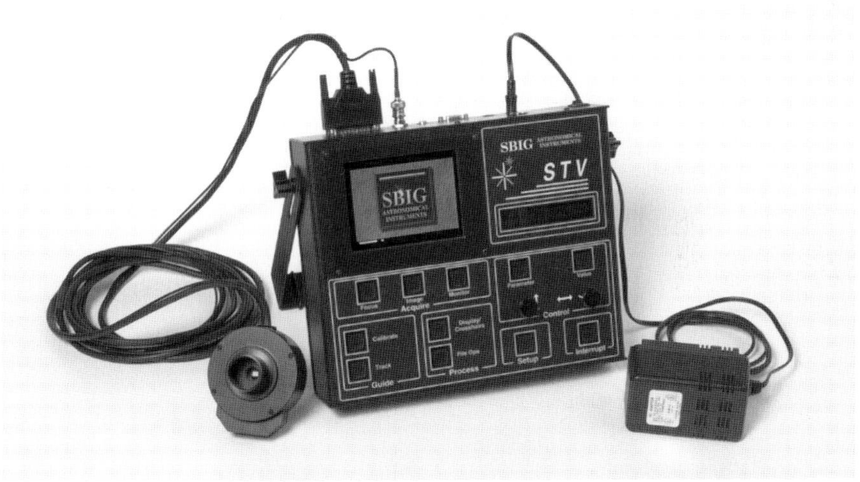

per second, considerably slower than the 25 to 30 frames per second of conventional video systems. Consequently, the STV's display exhibits some flickering and doesn't appear quite as seamless.

FIGURE 3.11
The Santa Barbara Instrument Group's sophisticated and surprisingly affordable STV video camera, introduced in 1999.

At the time of writing, the STV system is considerably more expensive compared to the price of dedicated analog-video astronomical cameras ($1,995 versus \$250 to \$850). Furthermore, in recent years video manufacturers such as Watec and Mintron have introduced a new line of extremely sensitive, moderately priced (\$600 to \$800) cameras with onboard frame-integration capabilities. These cameras can produce dramatic deep-sky images with refresh rates of up to 2.6 seconds (see Chapter 14 – The Deep Sky).

As demand for more advanced video systems increases, the price gap separating analog and digital video cameras will continue to narrow down. Digital camera manufacturers have been continuously improving the price-to-performance ratio of their consumer products and are beginning to close the gap. Throughout this book our discussions will focus mainly on analog-video systems. While analog video will eventually share the same fate as that of slide rules and phonograph records, the principles and many of the techniques described in this book will continue to have enduring value.

Video Signal Processors

4

If you want to manipulate the images contained in an analog video camera's output signal, there are two ways you can go about it. A stand-alone video signal processor (Figure 4.1) will allow you to manipulate the analog video signal in much the same way that a photographer manipulates an image when printing a negative in the darkroom. Indeed, the similarities between the two are striking — in both cases it is possible to alter the brightness and contrast of an image. However, a video signal processor offers even more capabilities. It can accentuate the edges of features in an image and make stretches to bring out details in the gray, murky areas.

Alternatively, you can convert the analog video signal from a series of voltage impulses to a digital form that can be manipulated with a far greater degree of control using a computer's image-processing or graphic arts software.

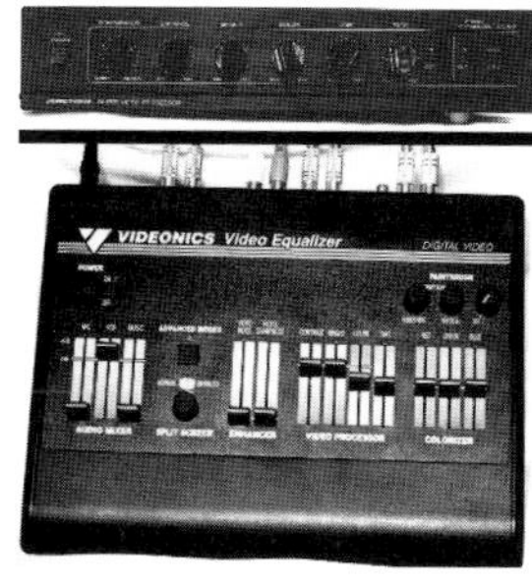

FIGURE 4.1

An analog video signal processor can greatly enhance the quality of images recorded on videotape but should not be used in conjunction with a computer.

A cautionary note is in order. If you intend to digitize your video images and manipulate them using a computer, a video signal processor will only prove superfluous and may actually be a handicap. Features accentuated by a video signal processor may look garishly unrealistic when subjected to digital image processing. In this chapter we discuss video signal processors and deal with computers in Chapter 8.

Brightness Control

This feature is essential, especially if your camera isn't equipped with a manual gain (sensitivity) control. Once you have adjusted the camera's shutter speed to get the brightness of the image approximately correct, it still may not be exactly where you want it. When videotaping the Moon, for example, you may want some part of a sequence to be perfectly exposed for a brilliantly sunlit crater and another sequence optimized for a dusky expanse of frozen lava closer to the terminator. With the brightness control you can optimize the brightness of various features as you watch them on your monitor.

More is at stake here than just getting a great-looking image. Another important factor is getting the best possible signal recorded on videotape. No matter how good the image may look on the monitor, if the camera doesn't provide a clean signal to the videotape, the result will be mediocre.

A clean signal begins with the camera. Underexposure produces a noisy signal and dim, grainy pictures. The video processor's brightness control can make the picture brighter, but the noise and poor definition will persist. No amount of tweaking can make an underexposed signal produce a good image. The solution in this case is to decrease the camera's shutter speed. If your camera doesn't have a user-adjustable shutter control, consider reducing the magnification.

Conversely, overexposure produces a white image with little if any detail. It can even cause the charge in one pixel to spill over into surrounding pixels, often called blooming. A video signal processor's brightness control can darken the picture, but the lack of detail will persist. Once again, no amount of tweaking can make an overexposed signal produce a good image. The solution is to increase the camera's shutter speed.

If your camera doesn't permit you to set its shutter speed manually and you're plagued by overexposure problems, you can make the best of a bad situation by stopping down the aperture of the telescope, though you'll reduce its resolving power, of course. If seeing conditions permit, you can always increase the magnification.

A third option is to insert a variable-density polarizing filter to reduce the amount of light reaching the chip. These inexpensive and readily available telescope accessories consist of a pair of polarizing filters that can be rotated with respect to one another. Each filter acts like a miniature picket fence, permitting only light waves vibrating in the direction of the pickets to pass. If the axes of the pair of polarizers are oriented so that they are parallel to one another, the amount of light transmitted is maximized, but if one polarizer is rotated with respect to the other so that their axes are perpendicular, virtually no light is transmitted. You'll be able to continuously adjust the amount of light falling on the chip over a very broad range.

From this discussion it should be obvious that a video processor's brightness control is reserved only for fine tuning. The main goal is to optimize the quality of the recorded signal.

Enhancement

The enhancement control allows you to increase the apparent sharpness of video images. After you've used a good enhancement feature, you may wonder how you ever managed to live without it.

To understand enhancement it is necessary to delve into the nature of the video signal itself. Composite video signals have two components — *luminance* and *chrominance.* Luminance consists of the black-and-white brightness and contrast (gray scale) components of the monochrome image. It is an amplitude-modulated signal that typically ranges from 0.0535 volts (black) to 0.714 volts (white). Chrominance contains the color information, which is absent in the signal of a black-and-white camera. Hue (dominant wavelength) and saturation (color purity) are replicated by combining the luminance information of three primary colors.

In the video signal processor (and in a VCR or monitor as well), these components of the signal are separated and sent down different pathways. When you "edge enhance" an image you outline the details in the luminance component with a thin, contrasting line, resulting in features with harder-looking edges. This technique is called *aperture correction.* It doesn't improve resolution, but it can make some elements of the picture stand out far more boldly and do it so impressively that you'll think the enhanced picture was taken with a different camera.

The human eye doesn't have a uniform, linear sensitivity to all patterns. By emphasizing certain frequencies in a video signal, images can be made to appear sharper or more pleasing. Boosting the signal response at around 200 lines is another common method of sharpening an image.

The problem with enhancement is that when you enhance the signal you also enhance the noise. However, this tradeoff can often be well worth it. If your signal is fairly clean, the noise will probably remain unobtrusive.

"Stretching" the Contrast of an Image

Stretching contrast is another common way to improve an image. An 8-bit black-and-white analog video signal contains 256 levels of gray. A video signal processor can stretch the upper midtone gray values into the white region and the lower midtone gray values into the black region. This can turn a so-so image full of murky grays turn into an eye-grabbing, spine-tingling scene. This is especially true in astronomy because many of the features on Mars, Jupiter, and Saturn seem to live in those middle gray tones. Contrast stretches are also excellent for bringing out rays on the full Moon and faculae and photospheric granulation on the Sun. There are limitations, however. Excessive contrast stretching can cause images to "posterize" by removing too many of the middle gray tones, resulting in a picture that looks more like a topographical contour map than an astronomical object (Figure 4.2).

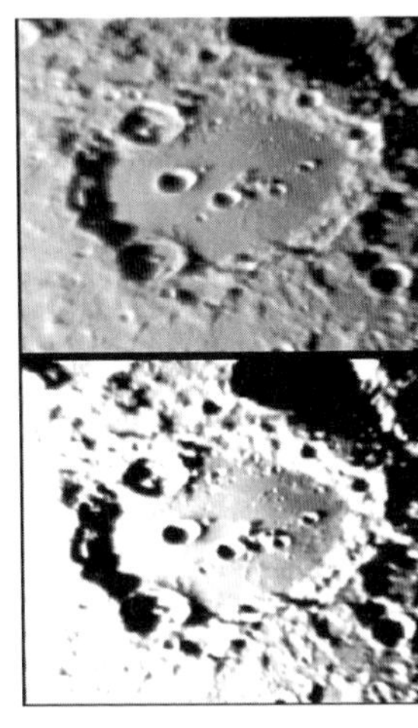

FIGURE 4.2

Excessive contrast stretching has been performed on the image of Clavius at bottom. The midtone grays have been forced into the black or white regions, producing a "posterized," or stepped, appearance.

Gamma Adjustment

Other kinds of stretching algorithms may be bundled with your video processor. One of these is the Gamma stretch. (On video processors this function is usually simply called Gamma. It should not be confused with the gamma control on your camera, which really is a contrast control.) Here the lower end grays are preferentially pulled out and the upper end whites are compressed. This is useful for images where features are buried in the lower end grays, not uncommon in images of the lunar maria. Gamma stretching should be regarded as a useful supplement rather than a substitute for contrast stretching.

Additional Features

Video processors are usually bundled with a variety of other features. Most are unnecessary and some are downright useless.

- **Color correction knobs.** If you use a color camera, these might just be worth having, but if you work in black and white you don't need them.
- **Time base corrector.** A time base error produces a "flag waving" effect in the lines of the image. Most time base errors encountered in video astronomy are caused by dirt on one of the VCR's recording heads or an old recording tape that has started to stretch. If you periodically clean your VCR's recording heads and use only high-quality videocassettes, you'll manage just fine without this feature.
- **Dropout compensator.** A dropout is a white line that flashes across the screen when the tape is played. Dropouts are caused by microscopic bald spots in the layer of magnetic particles coating the videotape. All tapes have them, but low-quality tapes have far more. The dropout compensator artificially fills in dropouts with an average of the brightness of the preceding and following lines. This feature is of cosmetic value only and is not worth spending money on unless you're making commercial, broadcast-quality videos. If dropouts are a recurring problem, buy better videocassettes. Videocassette quality is discussed in Chapter 7.
- **Noise reduction.** Noise-reduction circuits work by a variety of means, but they all smooth out the peaks in the luminance part of the signal. Unfortunately, those peaks usually contain many of the details in an image. When you use a noise-reduction circuit, objects tend to lose their sharp edges. The effect of many noise-reduction controls is subtle at best and can often only be clearly discerned with the aid of a split-screen display.
- **Editing equipment.** Processors frequently bundle in editing features just in case you need to prepare a video presentation. These include odd-sounding functions like "fades" and "wipes." If you give presentations to the public, these may useful, but most of us are concerned only with improving image quality.

Choosing a VCR 5

Now we need to discuss recording the images captured by the video camera. Rather than making a photographic print that can never be altered, video images are stored on videotape or on a computer's hard drive and retrieved for modification and printing at a later time. The ability to easily and repeatedly manipulate images is one of the greatest advantages of electronic imaging. The storage of video signals directly to a computer's hard drive is discussed in Chapter 8. However, since it's far more likely that you'll want to record on videotape, we'll talk about that first.

A videocassette contains a long, thin strip of plastic tape coated with a layer of magnetic material. Inside a VCR, motorized rollers move the tape at a constant speed past electromagnets known as *heads*. The current of the input video signal passes through windings around the cores of the recording heads, inducing magnetic fields. The density of these magnetic lines of force is determined by the magnitude of the video signal current. When the signal is recorded, each recording head lays down a diagonal stripe or *track* of alternating north-south magnetic poles in the magnetic layer of the tape. This process is called *helical scan recording*.

The reverse process occurs when the tape is played back. The magnetic lines of force from the pattern on the tape pass into the cores of the playback heads, inducing voltage impulses in their windings that are proportional to the strength of the magnetic fields, to replicate the original input signal. When a tape is erased, a powerful electromagnet demagnetizes the tape, wiping out all the stored information.

VCRs are complex marvels of mass production that employ components made to exceedingly precise mechanical tolerances. The narrow tracks required for high recording densities require that the tips of the magnetic poles of the video heads be very narrow

— only 0.001 to 0.005 inch (0.025 to 0.125 mm). By comparison, a typical shaft of human hair has a diameter of 0.0016 inch (0.04 mm)! Decades of research and development efforts have resulted in improvements in the design and fabrication of the record and playback heads. Combined with parallel improvements in videotape, these advances have contributed to steady progress in VCR performance, from the VHS format to Super-VHS (S-VHS) and from 8-mm to Hi8.

FIGURE 5.1

A four-head VHS-format VCR is suitable for recording and playing back astronomical images, but an S-VHS or Hi8 VCR will markedly improve picture quality.

Recording Formats: VHS or S-VHS? 8-mm, Hi8, or MiniDV?

You probably want to use the VCR that you already own. After all, why spend the extra money? Most household VCRs are VHS format. (VHS stands for Video Home System.) Unfortunately, this is a low-resolution format. Your images may look fine on your television set, but it's a low-resolution device too! The definition and sharpness of your images may prove quite disappointing if you make hard copies, so let's look at this in more depth.

If your video camera delivers more than 240 TV lines of resolution, you have more information in your output video signal than a VHS-format VCR can record. The same is true of the 8-mm format of many camcorders. Both are fine for use with the TV set in your living room because in the United States TV sets provide only 240-250 lines of horizontal resolution when receiving broadcast signals. (Their resolution increases to 340 lines when they serve as a closed-circuit monitor for signals fed to their video input jack.)

If you want to take full advantage of your camera's

superior resolution, you'll need a VCR that records in the S-VHS format or the Hi8 format of many camcorders. Both are capable of recording 400 to 420 TV lines of resolution. The difference in performance isn't just a subtle abstraction. Viewed in VHS format an actor's skin often blends into a uniform beige, but with S-VHS you can see freckles. If you record an identical high-resolution (>400 lines) video signal on both a VHS and an S-VHS VCR and print the results using a high-resolution inkjet printer, the superiority of the S-VHS recorder will be glaringly obvious. However, if your camera provides only 240 TV lines of resolution or you only intend to view your recordings on the TV set in your living room, don't waste your money on an S-VHS or Hi8 recorder.

FIGURE 5.2

A comparison of the recording resolution of VHS (center panel), S-VHS (at right), and Betacam SP (far right). Using a 70-mm Tele Vue refractor equipped with an Astrovid 2000 video camera and a Wratten #25 red filter, Ron Dantowitz captured this series of images of a dollar bill from a distance of 4.6 meters. The finest details appear in the Betacam SP image, with S-VHS a surprisingly close second.

The extraordinary images by Ron Dantowitz that appear throughout this book were recorded using a professional, studio-grade Betacam SP-format tape deck, but he readily admits that a far less expensive S-VHS or Hi8 VCR would be a viable alternative. Sony's Betacam SP format is used by the majority of electronic news-gathering organizations. Most broadcast stations demand, or at least strongly prefer, Betacam SP source footage, and many clients of professional video-production firms specify Betacam SP for industrial shoots.

The relative performance of the Betacam SP, S-VHS, and VHS formats is illustrated in Figure 5.2, which shows the fine details of a $1 bill imaged through a 70-mm Tele Vue refractor with an Astrovid 2000 camera from a distance of 15 feet. The clarity of the S-VHS image is considerably better than the VHS image and is not markedly inferior to the Betacam SP format. In fact, it looks very much like the Betacam SP image viewed through a very thin film of wax paper.

Dantowitz attributes the marginally superior image quality of Betacam SP to better recording fidelity at the high frequencies that correspond to the most closely spaced contrast changes in the image. By applying a weak low-pass filter with a video signal processor or an image-processing algorithm, he can make the Betacam SP image indistinguishable from its S-VHS counterpart.

Granted, Betacam SP is the best, but an S-VHS recorder is no less than an order of magnitude cheaper! Another consideration is the cost of the videocassettes. At $40 for a 90-minute Betacam SP tape, the price is downright exorbitant compared to an $8 top-of-the-line 120-minute S-VHS cassette.

Listed in order of decreasing image quality, the various video formats are: Betacam SP; S-VHS and Hi8; U-Matic ¾-inch; VHS. Although the exact order of this list could be debated, Betacam SP is definitely the best, with S-VHS and Hi8 very close runners-up at a small fraction of the cost.

For those of you who may be tempted to record your images using a digital video camera or recording device, beware! Most digital recording decks use storage algorithms that compress image data so it takes up less space on the tape. Unfortunately, these compression algorithms can introduce errors in the data, which can show up as spurious details in the still images. These artifacts could pollute your otherwise scientifically useful data, so use any digital recorder with extreme caution.

By the way, if you already own an S-VHS or Hi8-format camcorder, it's probably equipped with a jack to accept video input from an external source, so it can be used to record the output of your astronomical video camera. Far less bulky than a stand-alone VCR and powered by rechargeable batteries, a camcorder's integral videocassette recorder offers a very convenient way to record video observing sessions with portable equipment at remote observing sites.

In the portable digital-video revolution, Sony's GV-D900 records/plays back in MiniDV format, which provides up to 500 lines of horizontal TV resolution. This is three times better than that of its analog counterparts, and thus produces a significantly higher signal-to-noise ratio. (The benefit of such digital-based system becomes readily apparent when you compare the noise level present in the picture with the playback paused on a single frame.) The GV-D900 has a 5½-inch color LCD screen that utilizes the same active-matrix technology used in laptop PCs for superior clarity, sharpness, and color accuracy. The Firewire/IEEE 1394 interface of today's MiniDV cameras allows you to download the already-digitized pictures from the camera to a PC with great speed and efficiency.

Y/C Separation

This term describes how video devices transmit and process signals. The video signal from a camera arrives at a VCR after passing through a cable. This cable carries the two signal components that we discussed previously — luminance and chrominance. Luminance is the Y part of the signal and carries the monochrome brightness, contrast, and line definition information. Chrominance is the C part of the signal and carries any color information. In VHS technology, the Y and C signals are combined, a process called *modulating the signal.* When a modulated signal arrives at a VHS-format VCR, it must first separate the Y and C components, a process called *demodulating the signal.* After the VCR does its recording, the signal is modulated again for transmission to the monitor or printer, which in turn must again demodulate the signal. This process of modulation and demodulation takes its toll on the signal-to-noise ratio.

Years ago video engineers developed a way to get around this problem. By keeping the two signals separate (Y/C separation) both in the equipment and in the cables that carry the signal, the deterioration in the signal-to-noise ratio caused by modulation and demodulation can be avoided. The result is a cleaner signal. Y/C separation is a feature of the S-VHS and Hi8 formats, but not VHS.

Signals and Noise

A VCR will unavoidably introduce some noise into the signal. As a rule of thumb, the more you spend on a VCR the higher the quality of its components. That usually translates into a better S/N ratio. However, many models in any manufacturer's line of VCRs share common components. The only differences between them are often options like timing and search features, which don't affect the quality of the signal at all. Don't assume that just because a VCR is more expensive it has less noise. The only way to reliably tell is to check the numbers. S-VHS sets employ higher-quality components and have higher S/N ratios.

Tape speed is another important factor. VCRs offer a variety of speeds through which the tape is fed across the recording heads. The slower speeds (LP for Long Play and EP for Extended Play) prolong the recording time of a videotape, but only at the expense of picture clarity. Prolonged recording times are accomplished by increasing the width of the recording tracks, which causes cross-talk (information leakage between tracks) and loss of signal. You work hard to get a good signal, so don't lose it by using slower tape speeds. Record only in the standard play (SP) mode.

VCR Options

VCRs offer a vast array of optional features, especially as you get into the higher-priced models. These options include search, timing, and repeating functions. The vast majority are unnecessary and whenever possible you should buy them only if you really need them. However, a few options deserve special comment.

The majority of VCRs are designated *HQ,* which stands for "high quality." This is well worth having. It basically means that the recorder has what video engineers call a low-emphasis enhancement circuit, and that's good. A VCR that isn't designated as HQ is probably made with components of lower quality and should be avoided.

Some VCRs have a *search function* to quickly locate the beginning of each new recorded segment on a videotape. This is a useful luxury.

Pricey VCRs often bundle in a *time base corrector.* This feature can be useful when editing tapes but is quite unnecessary for most astronomical applications.

Some VCRs have *digital auto tracking.* This is a worthwhile feature. Many VCRs have manual tracking controls as well, and this too is useful. If your recording heads get dirty and start recording at a slightly slower speed, with manual tracking control you will be able to salvage the affected recording.

An integral *noise-reduction circuit* may sound like a nice feature but may subtly decrease the edge sharpness of images. Unfortunately, this is a common fixed function in many of the better VCRs and you may just have to live with it.

Finally, a *jog/shuttle control* lets you advance or reverse a videotape while examining it a single frame at a time. This feature is invaluable if you plan to capture and composite high resolution images.

Caring for Your VCR

There are a few things you should know about the care and feeding of VCRs. Follow a few simple procedures regularly and you'll increase the useful lifetime of your equipment.

Over time it's all but inevitable that the recording and playback heads in your VCR will accumulate motes of extraneous dust and particles of magnetic material shed by videotapes. The symptoms of dirty heads are *snow* (visual noise) on tapes you record or playback, or a rolling picture accompanied by other tracking errors. If these symptoms appear, buy a dry head-cleaning cassette. These look like ordinary videocassettes, but the tape they contain is actually a mildly abrasive scrubbing pad. They're simple to use and safe for the VCR when used as directed. A head-cleaning cassette usually contains enough tape for a 15- to 30-second pass over the recording/playback heads. After inserting the cassette and pressing the Play button, the tape will stop once it reaches the end. Be sure to follow the manufacturer's instructions — excessive use may damage the heads. (Note: Many VCRs of recent manufacture are equipped with a built-in head-cleaning mechanism. If your VCR is so equipped, you won't need to use a head-cleaning tape as frequently, if at all.)

If you are familiar with the internal workings of your VCR and feel confident enough to venture inside, the most effective way to clean dirty heads is with a cotton swab moistened with a frugal amount of head-cleaning solution. Never touch the silver cylindrical drumhead with bare fingers. You risk contaminating it with body oils that can markedly degrade your VCR's performance.

Keep your videocassettes neat and tidy. Store them in the dust jackets they came in. If they do show a visible accumulation of dust, wipe them off with a lint-free cloth before inserting them in your VCR.

Don't try to save shelf space by propping your VCR on its side. It's designed to run on a flat surface. Store your VCR in a cool, dry place to prevent moisture condensing on its internal components. The temperature and humidity cycling that occurs in most observatories can wreak havoc.

If you protect your VCR with a dust cover, always remove it before powering up the VCR. The VCR's vents need unrestricted air flow to properly cool its internal components for optimum performance.

Video Monitors 6

You'll spend some of your most enjoyable moments in front of your monitor looking at the images you record. While videotaping, you can seldom devote your full attention to the subject matter — you're focused on the technical aspects of getting a superior recording. The payoff comes later on, alone in your study in front of the screen.

So how do video monitors work? The electrical impulses of the video signal are converted into patterns of light to make a picture. An electron gun in the picture tube emits a scanning beam of electrons, spraying the screen from left to right with a succession of horizontal lines, starting at the top and moving down (Figure 6.2). The screen is coated with a fluorescent chemical or *phosphor* that glows when it is struck by electrons. As the scanning beam rapidly traces the image, the intensity with which the phosphors glow is proportional to the number of electrons that strike them. The stronger the beam, the brighter the light that is emitted on the screen.

FIGURE 6.1
A video monitor can be a standard television set or a dedicated, closed-circuit, high-resolution device without a tuner for selecting channels, like this one.

The series of lines traced by the electron gun are known as *scan lines.* Even though the electron gun "paints" the picture very quickly, in early television sets by the time the beam reached the bottom of the screen, the image at the top of the screen had started to fade. The solution was to subdivide the picture into numbered horizontal strips, projecting the odd-numbered strips first, then the even-numbered ones.

The odd- and even-numbered strips (called *rasters* and *horizontal retraces,* respectively) pair up to make a full picture or video frame. In the monitor the first field is converted into the odd-numbered scan lines, the second field into the even-numbered ones. When the electron gun sprays an interleaved pattern of fields, it sprays the first field from top to bottom, then it returns to the top of the screen and sprays the second field. This system is called *interlaced scanning.* The two fields are scanned either 1⁄50 second apart (with the European signal format, PAL, the acronym for *Phase-Alternating Line*) or 1⁄60 second apart (with the North American and Japanese signal format, which is called NTSC for *National Television Systems Committee*). Since two fields make up each frame, there are 30 frames per second in NTSC TV and 25 frames per second in PAL TV.

These video-signal formats scan either 525 lines (NTSC) or 625 lines (PAL). Of the 525 scan lines of the NTSC format, only 480 lines contain picture information; of the 625 scan lines of the PAL format, only 576 contain picture information. The other lines are used for internal timing.

Color video monitors employ three electron guns: one for red, another for green, and a third for blue. The phosphors are clustered on the surface of the screen in groups of three, one phosphor group for the electrons emitted by each of the three guns. A shadow mask behind the screen confines each beam to its corresponding phosphors. The tiny phosphors are bunched together so closely that the three primary colors they emit blend to give a full-color picture.

The resolution of a television or video monitor is expressed in TV lines. A monitor with more TV lines is the preferred option, provided they're not wasted on a camera or VCR of lower resolution. In the United States, home televisions can provide 340 lines of horizontal resolution. European televisions can provide 409 lines. The output of your video camera is known as a *composite* video signal, the same type of signal that a television station uses as a broadcast signal. Before this signal enters your home, it has been converted to a radio frequency signal of lower resolution, so when you watch broadcast television shows you see only about 240 lines of resolution. But when you connect a video camera or VCR to your television by using the video input jack rather than the antenna input, you can take advantage of the higher-quality signal and achieve 340-line performance.

Video monitors should not be confused with computer displays because they have different signal formats — analog (NTSC or PAL) for video and digital (VGA or SVGA) for computers. Analog video technology is based on the interlacing of scan lines, but an SVGA computer monitor paints the complete image in a single pass. The proposed High Definition TV standard also capitalizes on advances in electronics and uses a non-interlaced display.

There are far too many television sets and VCRs in use worldwide for NTSC and PAL to simply be abandoned anytime soon. Consequently, the consumer digital video systems that are being introduced retain NTSC and PAL as optional signal formats.

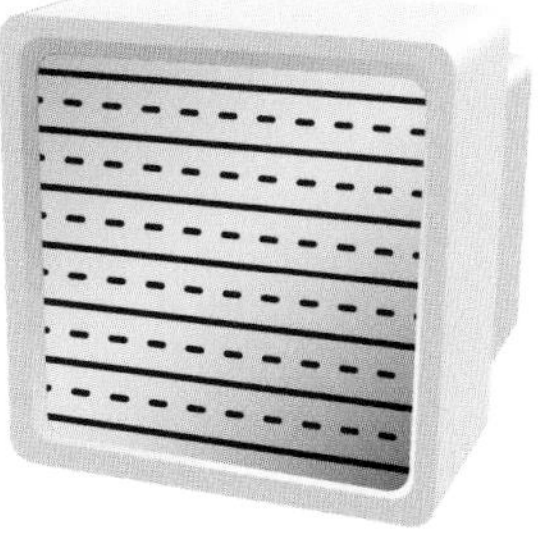

FIGURE 6.2
The electron gun located at the rear of a video monitor's vacuum tube produces odd and even scan lines. Each odd or even field, scanned in 1/50 second in the European PAL format with 25 frames per second or in 1/60 second in North American and Japanese NTSC format with 30 frames per second, represents only half of the complete image. Combining the odd and even scan lines produces a complete video frame.

TV Set or Monitor?

A TV set used to be synonymous with a picture tube attached to a tuner for selecting channels. A monitor displayed images but had no tuner and was used in the closed-circuit mode. With advances in video technology these distinctions are beginning to break down. To make matters even more confusing, in the past most color monitors were of lower resolution than black-and-white monitors, but even this has changed.

Amid all this confusion, what do you need and how do you find it? The answers to these four questions will guide you through the maze:

1. Do you need black-and-white or color? The answer to this question will be determined by your choice of camera.

2. TV or monitor? Do you need a tuner? Will you use the set to watch broadcast or cable channels or devote it exclusively to your astronomical pursuits?

3. Will you use the set for public demonstrations or just for personal viewing? If you make lots of public demonstrations, you'll want a large screen (19-inch diagonal or bigger). For personal viewing, a 9- to 14-inch screen size will perform admirably and be more portable. Give some consideration to how much weight you're willing to carry around if you have to lug the set back and forth from indoors to observe.

The confines of many amateur observatories are rather restricted, making a large screen undesirable because the observer is unable to achieve the proper viewing distance.

4. What kind of resolution do you need? If you have a high-resolution (>400 TV lines) camera, we urge you to purchase a high-resolution monitor to take full advantage of its performance. Not only will images be sharper and more detailed, you will find it easier to tell when the camera is in perfect focus.

Before you purchase a high-resolution video monitor, you should consider Leadtek Corporation's V-VGA300 Smart Video Converter (available from Stark Wholesale Electronics, 444 Franklin St., Worcester, MA 01604), a "black box" that permits any VGA or SVGA computer monitor to display NTSC or PAL video signals. This compact, inexpensive device connects to the output cable of your video camera or VCR, then plugs in between your computer and its monitor. A built-in switch allows you to select input from either the computer or the video source; no software is required. High-resolution computer monitors are currently manufactured in far greater numbers than high-resolution analog video monitors (particularly color units), so they tend to be far less expensive despite their superior performance. With the advent of this device, they have become the economical alternative!

Where Do I Put the Monitor in My System?

To record the best possible image on videotape, don't put anything in the path of the signal that doesn't need to be there! Your signal should go straight from the camera into the VCR (via a video signal processor if you use one) and only then to the monitor. Don't put the monitor ahead of the VCR, where it will only add avoidable electronic noise to your recordings. Remember to position your monitor in such a way that you can see it easily from the telescope when centering your subject and achieving focus.

Figures 6.3 and 6.4 illustrate some ways to set up your system.

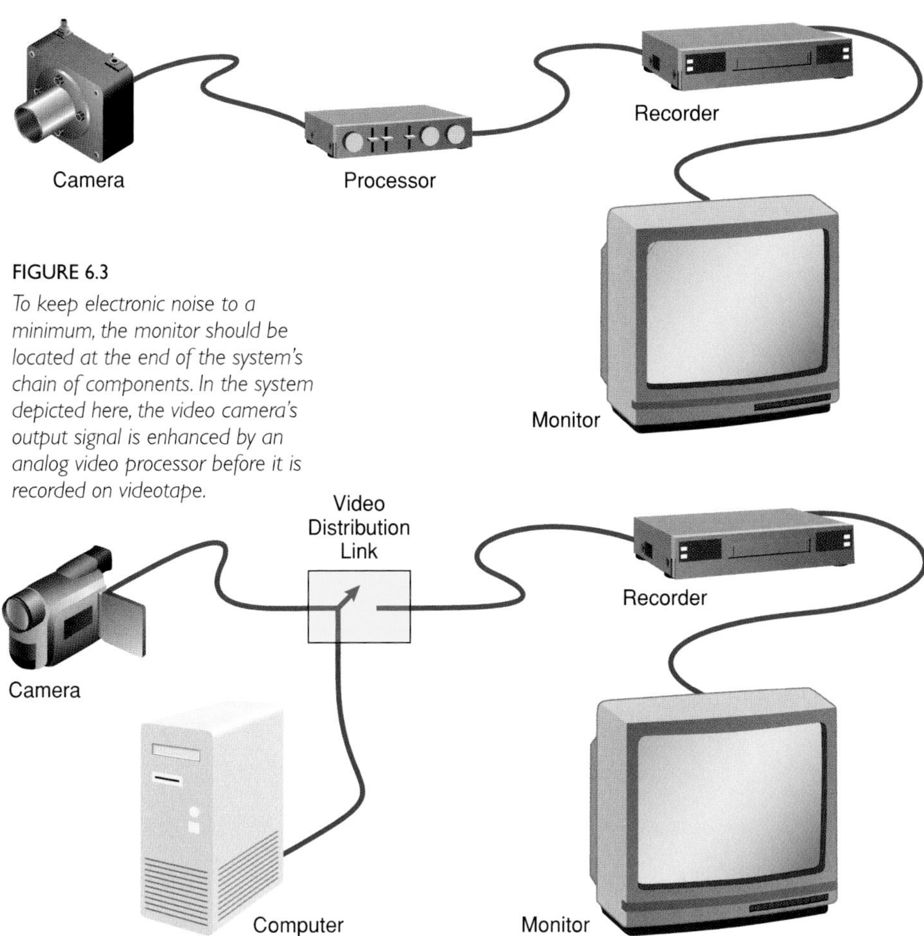

FIGURE 6.3

To keep electronic noise to a minimum, the monitor should be located at the end of the system's chain of components. In the system depicted here, the video camera's output signal is enhanced by an analog video processor before it is recorded on videotape.

FIGURE 6.4

In the system illustrated here the monitor is once again located at the end of the chain of components. The video signal is split via a physical switch or video distribution amplifier and fed to both a computer equipped with an image capture board and to a VCR. The VCR serves as a backup for any data losses that might occur in a computer-captured sequence.

Focusing with the Monitor

The faint halo produced by an out-of-focus star appears as midtone grays. A high setting on the monitor's contrast control will cause a faint point source to disappear when it is only slightly out of focus. Hence you may be very close to the correct focus setting but be unable to determine it. This lack of visual cues can be extremely frustrating for the beginner. Proper adjustment of the monitor will avoid this problem. Trim your monitor's contrast to about the midpoint between maximum and minimum and set the brightness control between middle and high. The screen's background should appear a dark charcoal gray rather than jet black. Now you'll be able to easily see the star's distended, out-of-focus halo and be able to judge at a glance just how far out of focus it is. Adjust the focuser to achieve the smallest point of light you can. Atmospheric turbulence may introduce confusing pulsations, so take your time. Once the image is in focus, restore the monitor's brightness and contrast settings as desired.

If your subject is extremely bright or the aperture of your telescope is very large, it is necessary to first check to make sure that the contrast control of the monitor is not at its maximum setting. This condition ruined some of Ron Dantowitz's best images of Saturn at Mount Wilson Observatory in 1997 when he recorded the planet through the 100-inch Hooker reflector. Saturn looked slightly burned out on the monitor, so Dantowitz adjusted the camera's gamma (contrast) setting. The following day he was bitterly disappointed when he played back the videotape! Saturn was dull, flat, and utterly lacking in detail, but it wasn't the camera's fault. By setting the monitor's contrast knob to display a high-contrast image, he had set the camera's contrast control too low, inadvertently lowering the contrast of the images that he recorded. Even experts make mistakes! Remember, an image that appears burned out on the monitor does not necessarily mean your camera settings are wrong!

You can profit from Dantowitz's painful lesson by setting your monitor's contrast a little bit below its maximum value, then setting your camera's gain control to its lowest setting. Then adjust the camera's electronic shutter until the image does not appear burned out on the monitor. Finally, adjust the camera's gain and gamma settings to achieve the most pleasing image on the monitor.

Videocassettes, Cables & Connectors 7

Selecting and Caring for Videocassettes

When you shop for videocassettes, you'll find many brands and specifications to choose from. But all videotapes are not created equal. No matter what the salesperson may tell you, they're not "all about the same."

Videotape consists of a base of tough polymer film coated on one side with a layer of magnetic material and on the other side with a layer of electrically conductive material that prevents the buildup of static charges. Without this layer moving videotape would generate electrostatic charges, attracting particles of dust and dirt that would quickly degrade the performance of a VCR.

The plastic base film must provide structural integrity and dimensional stability with respect to temperature, humidity, and time. Higher-quality tapes have a thicker base to prevent stretching. Once a tape stretches, it creates timing errors that cause linear features in the image to appear to undulate. This annoying effect becomes more pronounced with continued use as the tape continues to degrade.

Beware of the tapes offered by a number of companies that record for 160, 180, or even 240 minutes in the standard play (SP) mode. It sounds great, but it isn't. Manufacturers can't simply wind 180 minutes of full-thickness, professional-grade tape onto the reels of a 120-minute cassette, so they reduce the thickness of the base layer. The more fragile, elastic tape in these products is not a worthwhile compromise for astronomical use.

On the topic of tape stretching, remember that the frequent viewing of a tape in your VCR's view fast-forward and view fast-reverse modes applies excessive tension to the tape, forcing it against the recording heads while it is in rapid motion. Frequent use of the pause command and the jog-shuttle control has the same effect. To prolong the life of your tapes use these controls as sparingly as possible.

Now, what about the magnetic layer where the image is actually recorded? It is composed of a magnetic powder carried in a binder that bonds it to the base film. The magnetic layer must be smooth enough to provide close contact with the recording head but sufficiently rough to maintain adequate friction for the tape transport mechanism. The preparation and deposition of this layer involves proprietary processes that differ from manufacturer to manufacturer.

The performance of the recording medium is determined by the composition, size, and packing density of the magnetic particles. Three types of magnetic materials are commonly employed: ferric oxide, chromium dioxide, and metal particles suspended in a lacquer to inhibit corrosion. Ferric oxide is comparitively inexpensive and suffices for most consumer applications that don't demand the ultimate in performance. Incorporating chromium dioxide provides recordings with signal-to-noise ratios 5 to 7 dB higher than those achieved with ferric oxide.

Metal particles provide even higher signal-to-noise ratios as well as improved frequency response. Frequency response is important because changes in the contrast of a scene scanned by a video camera produce varying electrical waveforms in the resulting analog signal. The more closely spaced the contrast changes, the higher the frequency of the corresponding waveform. Consequently, there is a close relationship between the bandwidth of the signal and the resolution of the image. To record fine details, a video system must have a larger bandwidth. As a rule of thumb, 80 TV lines of resolution are provided per megahertz (MHz) of bandwidth. A VHS-format signal that produces a resolution of 240 TV lines has an implied bandwidth of 3.0 MHz; an S-VHS signal that produces a 400-TV-line image has a bandwidth of 5.0 MHz. The Digital Video Cassette (DVC) format adopted in 1994 will deliver 500 lines of TV resolution, equivalent to a bandwidth of 6.25 MHz.

Both VHS and S-VHS videotapes are ½-inch (12.7-mm) wide. Likewise both 8-mm and Hi8 videotapes are, as their names imply, 8 mm (0.315 inches) wide. The fact that both VHS and 8-mm format videotape can record only 240 TV lines of resolution but S-VHS and Hi8 (based on corresponding tape and cassette dimensions) can record 400 TV lines is largely attributable to the improved formulations of their magnetic recording layers. Hi8 tape truly is a technological leap that relies upon either metal particles or a vacuum-deposited layer of evaporated metal (similar to the aluminum coating of a telescope mirror) to cram all that signal information into such a limited space.

When you go to the store, look carefully at the grades of tape available. You'll probably find three: standard grade, high-grade (also called extra high-grade), and pro-grade. Sometimes you will see a high-quality grade, too. Predictably, pro-grade is the most expensive. It has a thicker base and will stretch less with prolonged use.

So, what's in a brand name? In the case of videocassettes, quite a bit. The best videotape is marketed under premier brand names like BASF, Fuji, Maxell, Sony, Memorex, Polaroid, and TDK. You'll pay a modest premium for these tapes, but it's well worth it. Read the labels carefully. You may encounter counterfeit brands with names like BSAF, TKD, or Suny that are similar but not quite identical to the brands that they are trying to mimic.

When shopping for VHS-format videocassettes, look for the official VHS logo on the package. Although it's not an absolute guarantee of superiority, manufacturers cannot print it on the box without agreeing to comply with a series of quality-assurance protocols established by the firm that developed the format, JVC.

S-VHS videocassettes cost about twice as much as their VHS counterparts. Again, look for the official S-VHS logo on the package.

Bargain-basement tapes employ inferior magnetic materials with a lower packing density. They usually suffer from a larger number of dropouts and a poorer frequency response that results in some signal loss and a lower signal-to-noise ratio. As if these problems weren't enough, they don't hold a magnetic charge as long and are harder to imprint.

Finally, it's worth noting that the two most common problems that occur with videotapes — stretching and dropouts — occur more frequently at the very beginning and at the very end of a tape. Sticklers about quality take pains to leave 10 or 15 feet of a tape blank at the beginning and end.

Store your videocassettes in their protective sleeves with the long edge pointing downward. If cassettes are stored standing on their side, the tape is more prone to sag and stretch under its own weight over time. Places to avoid keeping your tapes include the top of your VCR and monitor, beside a radiator, in your car, or a damp cellar. Keep your videocassettes away from external magnetic fields at all times. Devices to avoid like the plague include audio speakers, microwave ovens, and older style telephones with mechanical ringers.

Cables

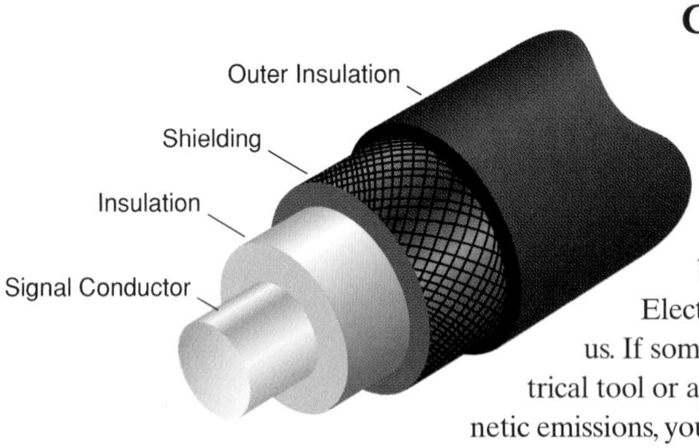

The main problem with cables is noise. In this instance the noise is not inherent in the cables themselves but comes from outside. Electronic noise is all around us. If someone nearby uses an electrical tool or appliance with electromagnetic emissions, you can often see the noise it generates as snow on your TV set or hear it as static on a radio. This kind of noise can also creep into your cables because a length of wire acts as an antenna.

FIGURE 7.1

A cross section of a shielded video cable. The signal is transmitted through a central copper wire that is isolated from external electronic noise by a surrounding shield of copper braid or foil connected to ground. These conductors are separated by a layer of insulation.

Shielded coaxial cables will reduce this problem. They carry a signal through a central conductor surrounded by a thick layer of insulation wrapped in a shield of braided metal or foil. The shielding intercepts outside noise and sends it to ground (Figure 7.1). While it's impossible to completely exclude noise, this arrangement can do a very good job.

The cables supplied with most video processors and VCRs are seldom of the highest quality. Where video cable is concerned, thicker is better. This isn't to say that cheaper cables won't work with your system; it's simply a matter of deciding how much you're willing to spend to have a really clean signal. We recommend RG-11U-type cable with an overall diameter of 13 mm, particularly if long cable runs are required. Miniature coaxial cables like type RG-174U are best avoided.

Keep your cables as short as possible. Longer lengths of cable make better antennas and gather more noise. Take pains to keep your cables as far away as possible from other electronic equipment and clear of the cord that supplies power to your telescope's motor drive or any other AC-powered equipment.

Wireless Video Transmitters

It is possible to replace video cables with a radio-frequency transmitter connected to the output of the camera and a tuned receiver located at the VCR, monitor, or computer. Wireless technology presents many advantages.

Eliminating cables adds a very welcome level of convenience. A wireless video transmitter offers the immediate benefits realized by untold thousands of musicians who were once anchored to their instruments and remote amplifiers by audio cables. Now with wireless equipment they can move around the stage free of cables. Likewise, a wireless video system means no cable between the camera and the recording device to trip over in the dark. The better (and predictably more expensive) units operate at very high frequencies (typically 900MHz to 6GHz (gigahertz)), so the more prevalent low-frequency ambient noise doesn't pose a problem. Any number of less expensive systems are capable of performing every bit as well, but they *are* more susceptible to low-frequency electromagnetic interference. The electromagnetic environment of your observing site will determine whether you can use them.

Connectors

The two most commonly used connectors for composite video signals are the RCA (push-on) and BNC (twist-on) types (Figure 7.2). You'll probably

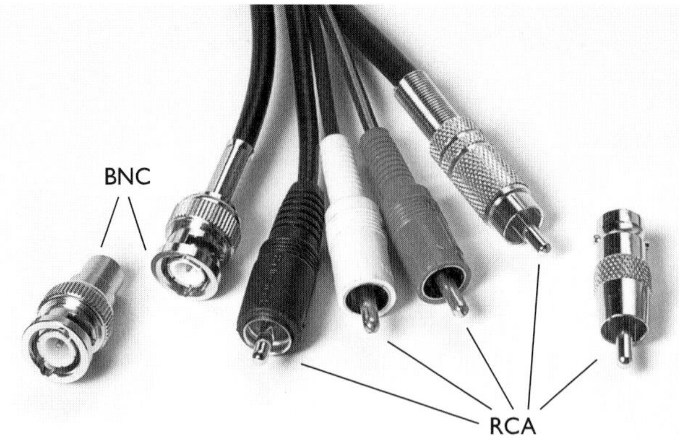

FIGURE 7.2

The most common connectors used with composite signal video devices are the BNC and RCA types.

encounter RCA connectors far more frequently these days, and they work quite well. Because they have a tendency to loosen over time push them in all the way and periodically check that they're still tight. Keep an eye on the collar to make sure that it fits snugly too.

Composite video formats like NTSC color and monochrome EIA are transmitted via a single channel and require only one coaxial cable. Many color cameras output two channels simultaneously, one for color information and one for brightness. Two-channel (Y/C) video devices (so-called *S-video*) use 4-pin DIN connectors. Each channel is carried by one pin and two pins connect to ground.

The main concern with video connectors is not their mechanical configuration but their materials of construction. Most metals corrode when exposed to moisture and atmospheric oxygen. Corrosion increases resistance to electrical current and reduces the strength and quality of a signal long before you can spot it with the naked eye. Gold is all but impervious to corrosion, even in harsh environments. Gold-plated connectors do cost a bit more, but they are well worth the investment.

Combining Video with Computers 8

Personal computers have revolutionized many aspects of our daily lives, including amateur astronomy. As with video equipment, the prices of personal computers have plummeted over the years and the technology continues to improve by leaps and bounds.

For video applications the personal computer offers an alternative recording mechanism to conventional VCRs. Encoding the output of a video camera onto a computer offers the immediate benefits of a single-step process for reduced noise. However, there are limitations to consider, especially compared to recording with videotape. But first let's discuss getting our images from the camera to the computer.

From Video Camera to PC

Transferring the image from the video camera to the computer requires a video-to-computer interface. These devices are known as frame grabbers, video graphics adapters, and frame capture boards. When a frame of video is selected or "grabbed," the analog signal is changed to digital by a device called an analog-to-digital (A/D) converter which samples the voltage signals and assigns numerical values to them. These quantized signals are then translated into a digital data stream, a process called *encoding.*

A digitized monochrome image consists of rows and columns of pixels, called lines and columns, respectively. Each pixel has an address that defines its location in an image and a numerical data value that quantifies its brightness. The number of discrete gray levels depends upon the internal electronics but is generally at minimum 8-bit. This means the brightness of any pixel is assigned one of 256 (2^8) discrete values. High-quality black-and-white frame grabbers are 10-bit devices that assign one of 1,024 (2^{10}) gray values or 12-bit devices that assign one of 4,096 (2^{12}) gray values.

A 24-bit color image is made by combining the 8-bit monochrome values of the three primary colors, red, green, and blue. This can be represented as 256 × 256 × 256 = 16 million colors. The range of tonal values or gray scale is still 256, however.

Digitized images can be imported into a graphic arts or image-editing program. Each frame grabber has its own unique control software. Some are more sophisticated than others. You can even move image files around your own software library, utilizing algorithms from different programs — each of which may offer something unique — to process them. In the next chapter, we will discuss sending image files to a printer to obtain hard copy.

FIGURE 8.1

External devices such as the Belkin USB 2.0 DVD Creator (top) and USB VideoBus II are low-cost solutions for connecting camcorders, VCRs, or other devices with standard video and audio outputs to your computer.

External Video-Capture Devices

Some years ago, external video frame-grabbing devices like Snappy and Snap Magic heralded a popular, low-cost means of capturing/converting analog video frames to digital files via the computer's printer (parallel) port. Unlike Snappy, Snap Magic features a "printer-through-port" connection, which makes it unnecessary to disconnect the printer from the PC when using the device. Snappy and Snap Magic both have RCA input and output jacks for connecting the computer and a video monitor. The resulting digitized frames can be read with any number of image-processing software.

Although these devices are still used by a few amateurs, they are quickly becoming extinct. They are now being replaced with products using USB 2.0 and the faster Firewire/IEEE 1394 interface standards. These "hot-swappable" interfaces allow you to connect or disconnect a multitude of peripheral devices without the need to restart the computer. All new PCs sold today come with one or more high-speed USB port connections. USB 2.0 boasts data-transmission speeds 40 times faster than the earlier USB 1.1 standard, which has a maximum data-transfer rate of only 480 megabits per second.

Today, a variety of economical USB-based external video-capture devices are available, which can operate under Windows and/or Macintosh operating systems. One example is the USB 2.0 DVD Creator package offered by Belkin. For less than $100, it can capture video and audio from webcams as well as camcorders, VCRs, and other analog-video sources (it's compatible with NTSC, PAL, and SECAM standards). It can capture high-resolution (DVD-quality) still images in 720-by-480-pixel NTSC or 720-by-576-pixel PAL format and videos up to 640-by-480-pixel resolution at 30 frames per second. Standard 640-by-480-pixel single-image snapshots to your computer are also possible.

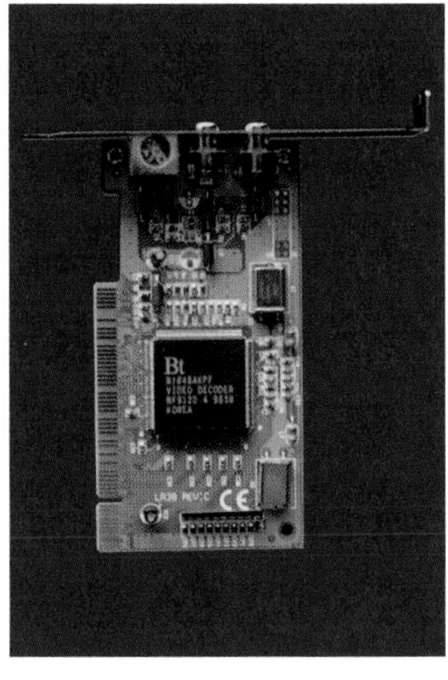

FIGURE 8.2

Integral video-capture cards offer an efficient way to capture full-motion video directly to a computer. Shown here is a simple, low-cost plug-and-play PCI slot card.

Video Graphics Cards

An option offered with many new computers today is the multifunctional, accelerated-performance video-graphics interface card. This internal card or integral hardware interface on the computer's main processor board drives the picture display on the computer screen. Some cards offer external video input/output connections for composite, or S-Video, devices. This allows you to display the computer's screen on another external monitor or capture images from another video source such as a camera or VCR.

Video-Capture Card Options

If your computer is not equipped for capturing video or it can't efficiently handle the workload when processing full-frame video, then you might consider purchasing one of the many specialized video-capture cards available. However, the hardware performance of these PCI slot cards, as with their bundled software, varies largely according to their purchase price. Plug-and-play cards range from basic models costing less than $100 to advanced units, such as the renowned Matrox Imaging-brand cards, which feature high-speed, real-time video codecs (compression and decompression algorithms) and independent onboard frame buffers and processing and can fetch thousands of dollars.

FIGURE 8.3

The rear of a computer with an installed video capture board showing the inputs for two RCA connectors and for a 4-pin S-VHS connector.

The ABCs of A/D Conversion

Bearing in mind your computer's existing processing capacity, to determine which frame grabber will satisfy your requirements you'll have to go back to the basics.

First, you need an adequate A/D converter. For example, when you perform a contrast stretch, the upper gray levels are being pulled into the white, and the lower gray levels into the black. Let's say you pull the upper 90 levels into white and the lower 90 into black. If you're using an 8-bit converter, this means your 8-bit (256 shades of gray) image is now living in the middle 76 levels (256 – 180 = 76). This may be a fraction of the gray scale you started with. Printing with such a small amount of gray values will usually give the image a posterized look.

You might imagine that if you started with a 10-bit A/D converter that provides 1,024 levels of gray, the loss of the upper 90 and lower 90 levels would still leave you with a whopping 844 levels. Unfortunately, the number of gray values that can be assigned to a signal from a video camera isn't simply a function of an A/D converter's capabilities. The noise inherent in a video camera's uncooled CCD array (its thermally generated dark current) and the noise of its supporting electronic circuitry limit the lower end of its dynamic range. Noise, which is expressed in electrons, is a measure of those electrons that are not generated by photons of light interacting with the CCD array.

There is also a limit on the number of electrons that can be accumulated in the pixels without saturating the chip before they can be shifted to registers and read out. This sets a limit to the upper end of the dynamic range. Video engineers refer to these parameters as a camera's *pixel depth* or *well depth,* which they often express as 8-bit depth or 10-bit depth. This value determines the corresponding number of gray levels that can be achieved by analog-to-digital conversion. A camera with a signal-to-noise ratio of 48dB is required for full 8-bit performance (256 levels of gray).

The ratio of well depth to noise defines the dynamic range of the CCD array. For example, a camera with a well depth of 2,600 electrons and 10 electrons of noise has a signal-to-noise ratio of 260:1, providing a little better than 8-bit performance.

Analog video cameras with 10-bit performance are available for very demanding industrial, scientific, and medical imaging applications, but they come with hefty price tags! It's a safe bet that any interline-transfer video camera that you paid less than a king's ransom for will not have the S/N ratio of 56dB required for 10-bit dynamic range. No 10-bit A/D converter can coax a 10-bit gray scale out of an 8-bit camera, so a good 8-bit A/D converter will serve your needs as well as they can be served, given the probable limitations of your camera.

Image Capture Software

Your frame grabber or video capture card will come with control software. Most software allows the user to set a variety of preferences before commencing a capture sequence. The typical preferences are:

- **Video Input Source Select.** These allow you to set the device to correspond with the input video signal format (generally Composite, RGB, and S-Video).
- **Frame Capture Size.** This allows you to select the window height and width (measured in pixels) of your still images or movies. The larger the capture window, the greater the demand placed on your computer. This can result in missing or dropped frames during capture sequences.

To ensure accurate, natural-looking images it is important to select a capture window with the 4:3 aspect ratio of the CCD chip in the video camera. For example, Jupiter's rapid rotation gives its globe an oblate, flattened appearance that is obvious at a glance even in a small telescope. If a square capture window is selected rather than the standard rectangle with a 4:3 width-to-height ratio, an unrealistic circular Jupiter will be the result.

- **Selectable Capture Rate.** This allows you to choose the number of frames to capture each second. Standard settings for PAL and NTSC frame rates of 1⁄50 and 1⁄60 second are included.
- **Capture Method.** This option allows capture to either the computer's random access memory

(RAM) or to its hard drive. If you have plenty of available RAM and you're capturing short sequences, selecting Capture to RAM is the most efficient method. RAM access is much faster than capturing to disk and provides the highest capture rates.

• **Compression Method.** This selects the compression ratio for the final images or movies. Compression is great for saving disk space by reducing the file size, but it irreversibly reduces image quality. It is highly recommended that the No Compression function be selected. However, with slower computers or when capturing larger frames, compression may be necessary. In this case, you should experiment to find a level of compression that does not appreciably affect image quality. Common options include 32-bit, 24-bit, 16-bit or 8-bit. Some capture cards also offer YUV 411, which takes the color information from your computer's virtual random access memory (VRAM) to produce color movies using a fraction of the disk space of standard 24-bit color.

• **Capture Length.** With this option a user can select a time setting for each capture sequence from one second upward.

• **Video Source.** As with a video signal processor, the bare necessities are the basic luminance and chrominance functions — brightness and contrast, hue and saturation. Contrast control allows you to stretch the murky middle grays. Brightness control allows you to alter the picture's exposure and get your target right in the middle of the gray regions. Neither of these functions improves spatial resolution, but both can pull details out of the murky gray tones characteristic of many lunar and planetary images.

Digital Video Formats

Your capture software will allow you to save the images in a variety of standard formats, and you should save files each time you alter them. Inexpensive frame grabbers offer the capability to capture only single frames, but specialized plug-and-play cards for PCs provide the ability to capture full-motion video in one of several formats, including MPEG-2. One of the most common digital video file types is called *Audio-Video Interleaved (AVI)* (Figure 8.4). Most capture software programs provide selectable compression modes to reduce the enormous file size of the finished sequence to manageable proportions, but image quality can suffer. Whenever possible use the uncompressed mode to get the most detail out of your final images.

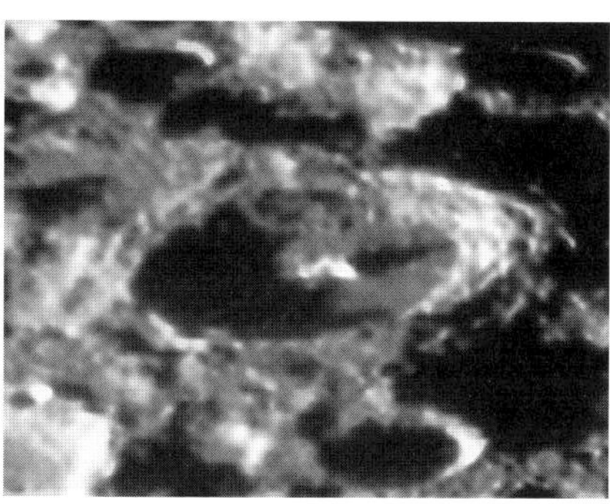

FIGURE 8.4
Moretus is a 114-km-wide crater located in the rugged southern highlands of the Moon. This image is one of only three sharp frames that occurred in an AVI sequence of more than 100 frames recorded under poor seeing conditions.

Which System?

At this point you may be wondering if a video processor, VCR, and monitor are really necessary! Again, this depends on your equipment and what you're trying to accomplish. Admittedly, feeding a video camera's signal directly to a computer equipped with a high-quality video interface board is the cleanest system with respect to electronic noise. But does your computer have enough processing power for real-time video capture? Do you really need to capture in real time? Perhaps you only want a few clean snapshots. Is your computer too cumbersome to carry outside? Can its hard drive tolerate a bitterly cold night? Let's examine these questions further to help you to decide what you need for your system.

Real-time Frame Capture. Real-time data capture places two demands on your system. First, your computer must be able to deal with truly enormous volumes of data. Video frames come in at a rate of 25 to 30 every second. In just 20 minutes, a rate of 30 frames per second will accumulate 36,000 frames! If you record for one hour, you're talking a whopping 108,000 images!

The second problem is that your video interface must deal with the fast rate of 30 frames per second (the *refresh rate*). Some of the more inexpensive models just can't handle that speed, so check the frame grabber's capabilities before you buy it.

In addition, your computer will need to have a high data-transfer rate for writing to your hard drive. If the data-transfer rate is inadequate, your capture software will report the number of frames dropped for that capture sequence. At 30 frames per second, your hard disk drive should have a minimum access time of 10 milliseconds and a data-transfer rate of 1.5 megabytes per second. If your computer's central processing unit (CPU) and interface can handle the data, you'll need only those two pieces of equipment. The computer monitor will perform all the functions of a video monitor.

If your computer can't handle the volume of data, use a VCR to record the images in conjunction with a conventional TV or dedicated high-resolution video monitor.

A final tip: When capturing images directly to the computer, close all nonessential programs that use up valuable memory and CPU processing power.

Capturing Single Frames. If you're only interested in obtaining a few stills of your subject then the demands on your computer are greatly reduced. You'll only need the video interface board and the computer (Figure 8.5). You can take several test shots to adjust the exposure, then capture frames until you grab one in which the effects of atmospheric turbulence are minimal. Bear in mind, however, that even the sharpest single frames will look grainy due to electronic noise.

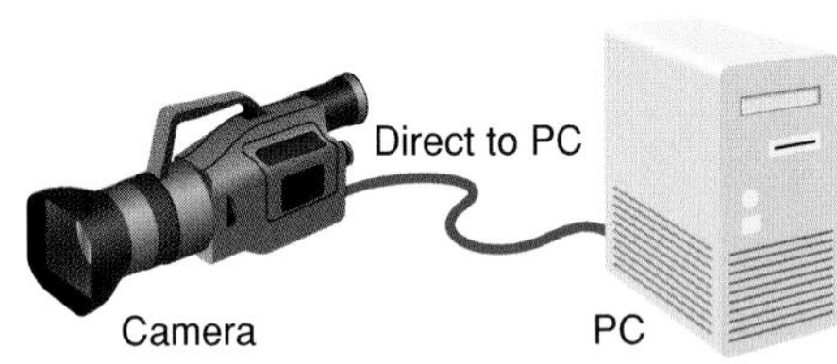

FIGURE 8.5

The most direct computer image capture system simply connects the output of a video camera directly to a frame-grabber board. The video display can be viewed on the computer's monitor.

Getting the Cleanest Signal. Whether your quest is single-frame imaging or real-time movies, always remember the rule for noise reduction: "Less is better!" Every additional piece of equipment introduces electronic noise, so use the fewest possible components.

The noisiest video capture system is camera to VCR to frame grabber to computer (with results shown in Figure 8.6). It should be noted, however, that the noise seen in video images is not always a direct result of equipment-laden systems such as this. The most common source of noise is from the use of poorly shielded cables or dirty, loose connectors. The second most common noise source is the radio-frequency interference generated by a motor running nearby — like the motor of the fan used to speed the thermal equilibration of a telescope's mirror.

Equipment Portability. If your computer is not portable, you may prefer to run a long video cable out to the telescope (Figure 8.7). Most cameras will manage with a cable up to 15 feet long without appreciable signal degradation. Remember, a long wire acts like an antenna and collects ambient electrical noise, so use heavily shielded cables. In addition to the video cable, you will also need to construct an extension to your telescope's drive controller so that minor tracking corrections can be made from the computer's location. Bear in mind that the hard drives of most computers perform poorly in subfreezing temperatures. You might also consider switching to a VCR.

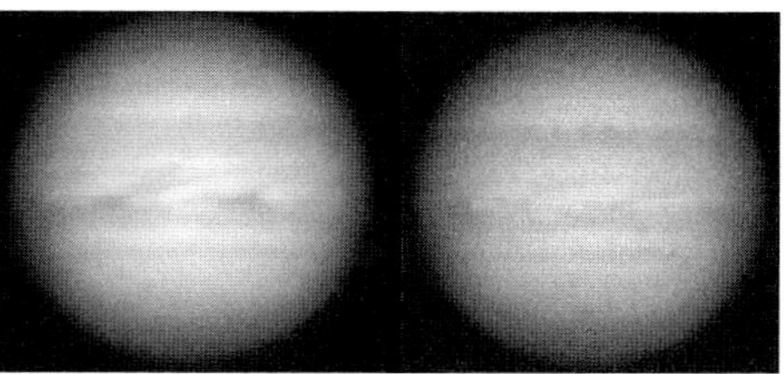

FIGURE 8.6

The single frame of Jupiter at left was obtained by capturing the output of an Astrovid 2000 video camera directly to a computer using a video capture card. The single frame at right was simultaneously recorded using an S-VHS VCR, then grabbed by the capture card when the videotape was played back. Despite the far greater demands placed on a computer's memory, capturing directly without recourse to videotape gives perceptibly superior images.

Digital Image Processing

Up to this point we have emphasized recording the best possible images. In this section we'll talk about processing the image to extract additional bits of data and improve the visibility of subtle details. While many image-processing programs offer a wide variety of special effects and complex functions, we'll focus on the most useful and frequently employed algorithms.

A digital black-and-white image consists of a series of numbers that define the brightness levels in the image. An 8-bit image has 256 gray scale values. The number for white is 255, which diminishes through ever-darker shades of gray until it reaches 0, the number for black. The various functions in your image-processing programs perform a wide variety of mathematical manipulations on these numbers to change the appearance of the image.

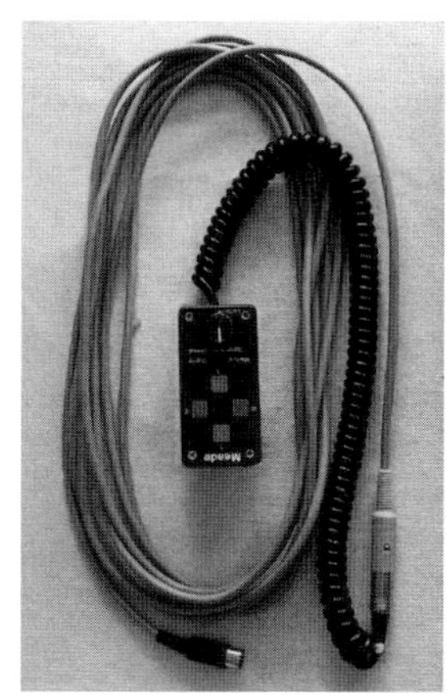

FIGURE 8.7

A hand-held controller of a commercial dual-axis telescope drive retrofitted with an extension cable to permit remote control of the telescope from an indoor computer workstation.

Low-Pass Filters. Low-pass filters smooth out the checkerboard pattern of pixels that makes up an image and reduce the objectionable stepped appearance of edge lines. They achieve these effects by decreasing the levels of contrast between adjacent pixels. The *average* algorithm combines the brightness value of each pixel with the brightness values of the eight surrounding pixels, then divides the pixel by the mean value of this sum. The gentler *blur* algorithm calculates the arithmetic mean of each pixel and four surrounding pixels. *Noise-reduction* algorithms are another example of low-pass filters, but they calculate the median rather than the mean values of each pixel and nearby pixels. All low-pass filters soften features in the image, just like throwing it out of focus. Sharpness can be restored by the subsequent use of complementary high-pass filter algorithms and by applying an unsharp mask.

High-Pass Filters. High-pass filters like *sharpen* increase the differences in brightness values of adjacent pixels but increase noise in the process. After an image has been smoothed using a low-pass filter, judicious use of a high-pass filter can restore much of the lost detail (Figure 8.8). The proper combination of low-pass and high-pass filters involves a compromise.

Unsharp Masking. This algorithm first creates a defocused image by calculating the average values of groups of pixels. With most programs, it is possible to choose the number of pixels to be averaged. This number corresponds to the degree of defocusing. The defocused image is then subtracted from the original image. This algorithm can be remarkably effective at accentuating small-scale, high-frequency details that would otherwise remain hidden. Use this extremely valuable function sparingly, however. Excessive unsharp masking will distort features and can even introduce spurious details that have no basis in reality.

Stretching an Image. Image-processing programs offer a variety of ways to stretch an image. The simplest, the contrast stretch, has already been described. This alters the slope of the linear relationship between gray scale input and gray scale output. In doing so it pulls the lower-end grays into the black and the upper-end grays into the white.

However, a linear gray scale may not be optimal for your image, especially if it's an image of a planet, so good image-processing software provides a variety of stretching functions. One is the *gamma stretch*.

Here the linear relationship between input and output is changed to a curve that slopes very steeply at first and then gradually flattens out. This has the effect of stretching out the darker grays in the shadows and compressing the bright upper-end whites in the highlights. Use this algorithm to accentuate features on the Moon's dark basalt plains. The gamma scale can also be inverted to compress shadows and extend highlights. This is particularly useful for accentuating features in the brighter regions of an image, like details in Jupiter's zones.

A third kind of stretch is called the *logarithmic stretch*. Here the lower-end grays are stretched even further and the upper-end whites compressed even more, smoothing out the kinks in a plot of the image's brightness values. This is a more severe stretch than the gamma stretch, so use it with caution.

Histograms. Not all stretches correspond to a linear or logarithmic function. Some programs allow you to construct a *histogram* of the image (Figure 8.9) and then modify it. A histogram is a graphical representation of the frequency with which the various gray levels occur in an image. The horizontal coordinate (x-axis) plots the range of brightness values (0–255 for the common 8-bit scale), while the vertical coordinate (y-axis) plots the number of pixels with a particular brightness value.

One useful algorithm is called *graphical image scaling*. Here the program spreads the image out across all the levels of gray without altering the shape of the histogram. This is particularly useful in images of relatively low contrast, where all the brightness values are clustered together. After spreading out the narrow range of values across the entire gray scale while maintaining the shape of the original histogram, contrast is increased without introducing distortion in the way that gamma stretching does.

Another useful histogram function is called *histogram equalization*. It is used to accentuate the visibility of faint objects. Here the program takes the histogram and forces an equal number of pixels into each brightness level, converting the histogram into a horizontal line.

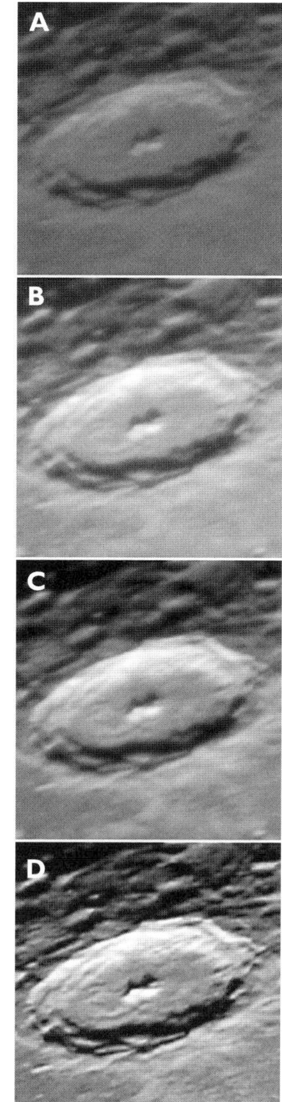

FIGURE 8.8

Beginning with a raw image (A), processing is used to improve the image. (B)The effects of a high-pass (sharpen) filter, (C) a low-pass (blur) filter, and (D) an unsharp mask on an image of the crater Langrenus.

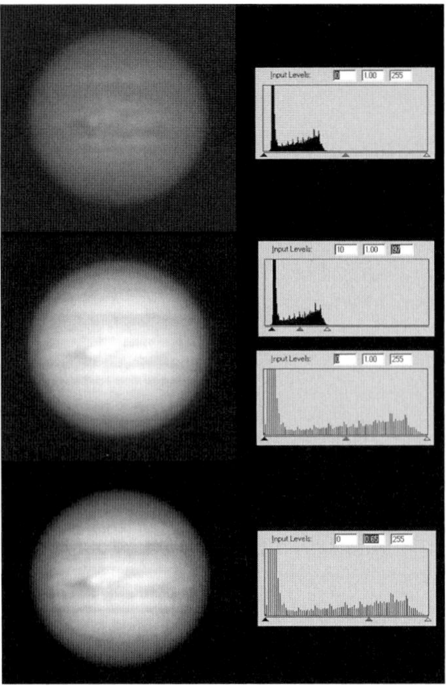

FIGURE 8.9

Image contrast can be improved with the histogram-stretch function of a program such as Adobe Photoshop. At top is a low-contrast image of Jupiter. Moving the sliders adjusts the brightness and contrast by remapping pixel values to higher (brighter) or lower (darker) numbers. Midtone adjustments help bring out detail in the most subtle features like the planet's belts, as shown in the bottom image.

Stacking Images. Image subtraction performed on pairs of images of various regions of the Moon taken through various spectral filters during the Apollo and Clementine missions greatly accentuated localized color differences to reveal variations in soil chemistry. Combining images is even more useful when recording the planets (Figures 8.10, 8.11). You'll find that single frames of the planets invariably have an objectionably grainy appearance caused by electronic noise. Remember, in addition to the electronic noise of its circuitry, the uncooled CCD array in a video camera generates considerable thermal dark current noise. The images produced by the uncooled array of a video camera have signal-to-noise ratios comparable to underexposed images made with a cooled CCD array camera. For all its many virtues, this is the single great failing of video. But by averaging a number of noisy frames together, it is possible to create a composite image with a vastly improved signal-to-noise ratio and aesthetic appeal.

Generations of astrophotographers have employed a very similar technique known as *composite printing* to suppress film grain. Since the locations of the grains in any set of photographic negatives are random and the number of grains is enormous, successively printing multiple negatives of the same subject (all in precise registration so that they are perfectly superimposed) onto the same piece of photographic paper will produce a print with much less obvious grain. Each negative is given an exposure in the photographic enlarger that is inversely proportional to the total number of negatives being composited.

Today you can mimic this laborious darkroom technique on your computer with digitized video frames. The improvement in the signal-to-noise ratio of the resulting composite image will be proportional to the square root of the number of images that you average. If four images are combined, the noise of the composite image will be cut in half; combining nine images will reduce it by two-thirds. Seasoned planetary observers who employ this method generally stack eight to 16 frames. It's a time-consuming exercise, but the results are well worth it if your goal is the best hard-copy image. All too often the greatest challenge is finding eight or more sharp images to stack!

Even in excerpts from a videotape recorded under very good seeing conditions, it is rare to find more than one really sharp image in every five if the tape is methodically reviewed one frame at a time. Moreover, two or more consecutive frames almost never appear equally sharp. The great advantage of converting analog video to a digital format is that it then becomes possible to composite nonconsecutive fields or frames that are chosen for their sharpness. Ron Dantowitz has coined the term *selective integration* to denote the process of compositing carefully selected nonconsecutive video fields or frames.

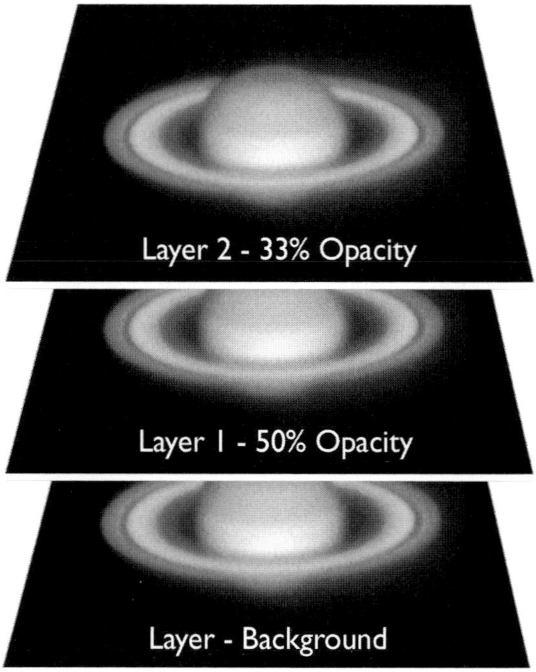

Layer 2 - 33% Opacity

Layer 1 - 50% Opacity

Layer - Background

FIGURE 8.10
Stacking frames to make a composite image of Saturn.

Selecting Images to Composite

Once you have captured dozens of video sequences with a capture device, the fun begins. Depending on the length and number of captured sequences, this exercise can get tedious. Imagine that at a rate of 30 frames per second you've captured 40 sequences, each 5 seconds long. That would be like sorting through 6,000 photographic prints! Even if your computer dropped 25 percent of

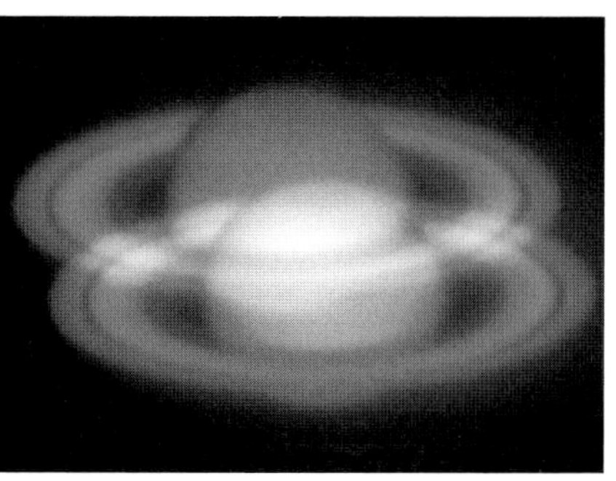

FIGURE 8.11
Two layers, each with 50 percent transparency, being brought into correct registration.

the images during capture, a whopping 4,500 frames would remain to be evaluated.

Your AVI playback software should allow frame-by-frame playback. Starting at the beginning of a movie file, step through each frame by using the right arrow key on the keyboard. (Reverse direction by using the left arrow key.) The best way to locate the most promising stills is to quickly tap the right arrow key as if you're transmitting a message in Morse code, pausing to make note of the frame numbers of the sharper images. Returning to the beginning of the file, you can now make a more critical inspection of these frames.

When reviewing a movie file of a crater-studded lunar landscape, keep an eye on the clarity of the entire image, not just a feature of special interest. Depending on how rapidly the seeing changed and the resulting level of distortion when the images were recorded, half of an image can often appear crisp while the other half is blurred or even appears warped.

Evaluating and selecting images requires intense concentration. If you intend to stack images to improve the signal-to-noise ratio and accentuate features, selecting the very best frames is imperative. If two frames are stacked or layered and one of the images is appreciably more degraded by seeing than the other, the result will be a composite image with less sharpness than the better of the two originals.

Once you have moved into the first section of the raw movie file previously noted for its clarity, the decision-making process begins. You will probably find about a half-dozen to a dozen single images worth keeping. So which is the best? This is where a keen eye is needed to carefully examine surrounding features. Pressing the arrow keys forward and backward to rapidly alternate between stills will reveal subtle differences in shapes of craters, mountains, and shadows caused by atmospheric turbulence.

Sometimes an excellent image will seem to jump out at you with unmistakable clarity when a minute craterlet or delicate rille seems to suddenly appear from nowhere. This might be the first frame to keep using the Capture to Clipboard or Save As File Type functions. That single image may be all you require. However, if you want to stack a series of still images, it sets a demanding precedent for the level of quality required from any images that follow.

Watch for very pronounced distortions of features within each image, even though they may appear to be in focus. For example, you might be examining a recording of a lunar mountain as it casts a long spire of shadow under early morning or late afternoon lighting. In one frame (we'll call it Image A) the shadow appears at right angles to the peak, but when you advance to the next frame, the angle of the shadow with respect to the peak has changed perceptibly. In addition, you notice that the shape of a nearby craterlet has also taken on a more pronounced elliptical form. By advancing through a few more frames of steady seeing, you determine that the shadow does indeed lie at a right angle to the peak and that the craterlet is in fact circular.

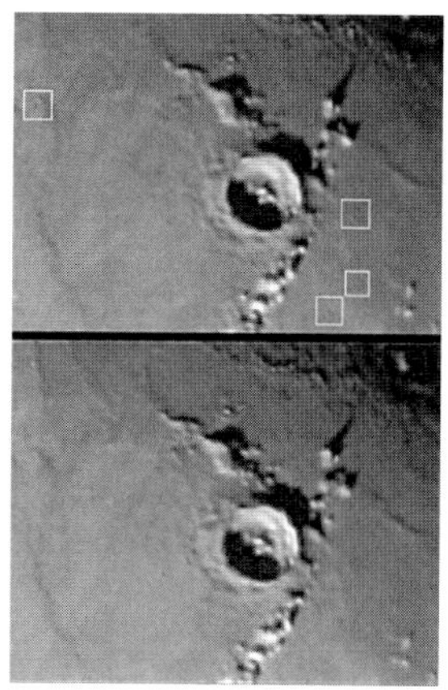

FIGURE 8.12
The effects of poor seeing often cause smaller craters to simply disappear. While the prominent crater Eratosthenes is relatively unchanged, subtle apparent changes occur in smaller features (enclosed in small boxes in the upper frame).

In this example, Image A can be saved, but the second frame represents a false impression and should be discarded. Another example of image distortion is the lensing effect caused by seeing cells passing through the field of view (Figure 8.12). When examining a succession of frames, you'll often notice that a planet or lunar crater suddenly swells, then quickly resumes its normal appearance. The effect is not unlike looking at the contortions of a living amoeba under a microscope. The trick is to identify and reject the images that contain spurious features that are merely artifacts of atmospheric turbulence.

This is another advantage of combining several fields or frames. Only real features show up repeatedly, while the random spurious features that may appear in individual frames are suppressed in composite images.

Dantowitz has devised a simple but extremely effective trick to locate the best images when he reviews his recordings. He readily admits that this technique sounds quite strange, but his results speak for themselves. Begin by taking a piece of stiff cardboard measuring about 8 by 11 inches. Cut a hole a little over 1 inch in diameter through its center. Take a

1.25-inch format eyepiece of about 25-mm focal length and shove it halfway through the hole in the cardboard. Now, impaled cardboard in hand, watch the video monitor though the eyepiece with its "eyeball side" facing the monitor. Hold the eyepiece and cardboard combination far enough from your eye that the image on the screen looks smaller, not bigger! Let the area of interest on the screen fill the reduced image you see in the eyepiece lens. You might imagine that the same effect could be achieved simply by stepping far enough away from the monitor, but there are subtle differences, chiefly that the cardboard cutout allows you to focus your attention on a small portion of the image exclusively without being distracted by the entire screen.

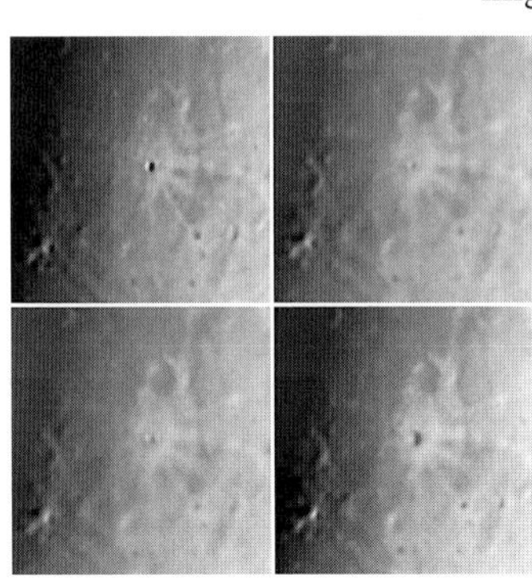

FIGURE 8.13

These four images show how dramatically atmospheric turbulence changed the appearance of the crater Kepler in four consecutive 1/25-second frames. The image at top left represents the average appearance. At top right, the crater's shadow-filled interior has lost contrast. At bottom left, the crater's shape is distorted, and contrast is reduced even further. At bottom right, contrast has been restored, but the apparent size of the crater has almost doubled due to atmospheric lensing.

Automated Image Stacking. Today, there are several commercial and free software tools that can automate the rather laborious task of manually stacking frames in order to improve the image's signal-to-noise ratio.

Three of the most popular programs originated from Europe and can process a sequence of standalone images or AVI media files. *AstroStack*, *Registax*, and *K3CCDTools* are widely used by amateur videographers for processing video, webcam, or conventional astronomical CCD images. *Registax* by Cor Berrevoets is an excellent, easy-to-use program that features powerful wavelet digital filters for enhancing stacked images. As with *Registax*, *K3CCDTools* by Peter Katreniak and *AstroStack* by Robert J. Stekelenburg also provide wizards that can, with some accuracy, automatically assess the quality of each frame selected for stacking. But, visual selection by the user is always recommended and, more often than not, produces the best results. With this in mind, each program allows the user to manually select the best frames to use for stacking and processing.

These programs can stack several hundred frames in a relatively short amount of time. The result is a composite with substantially reduced visible noise. The improved image detail means you'll obtain more effective results when you apply unsharp masking or other image-enhancement filters.

When using *AstroStack*, the user selects the start and end of a series of images (or loading an AVI file), the program stores each frame in the order that it occurs in the source file (e.g., "crater00.bmp," "crater01.bmp," "crater02.bmp," or in the case of an AVI file, frame 1, frame 2, frame 3).

Once loaded, the user can utilize the Page Up and Page Down keys to review each image and press the

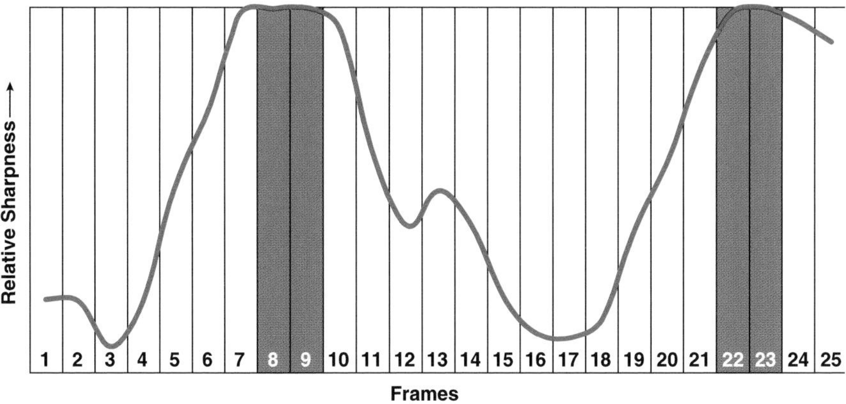

FIGURE 8.14

The sharpness of video frames is rated here on an arbitrary scale ranging from 0 (best) to 5 (worst) during a one-second capture sequence as a 10-inch Newtonian recorded Saturn at f/30 under turbulent seeing conditions. Only two of the 25 frames produced images of acceptable quality. Frames 22 and 23 were reasonably good, but detail in the rings was inferior to frames 8 and 9.

associated function keys to assign quality ratings. The quality ratings are: "don't," meaning don't use an image; "bad" for images blurred by poor seeing; "average" for images of average quality; and "good" for the sharpest images in the sequence. Clicking on the Calculate command causes the program to add the images by their corresponding pixel values multiplied by a quality coefficient, then divides the resulting values by the sum of the quality coefficients of each image. The resulting composite image is a weighted average that emphasizes the sharper images and de-emphasizes the poorer ones.

For example, let's say you have four frames to composite to which you've assigned the following quality ratings: frame 1 – don't; frame 2 – bad; frame 3 – average; frame 4 – good. The pixel value XY of the composite image would be calculated as follows:

Composite image's value at pixel XY = [0(frame 1's value at XY) + 10(frame 2's value at XY) +

20(frame 3's value at XY) + 40(frame 4's value at XY)] / (0+10+20+40). Frame 4 comprises 57 percent (40/70) of the composite image; frame 3 has 29 percent (20/70); and frame 2 contains 14 percent (10/70). Of course, frame 1, with a "quality coefficient" of 0, is not used at all.

If all four frames were assigned the same "average" quality rating, the pixel value of the composite image would be calculated as follows:

Composite image's value of pixel XY = [20(frame 1's value at XY) + 20(frame 2's value at XY) + 20(frame 3's value at XY) + 20(frame 4's value at XY)] / (20+20+20+20). This expression is equivalent to [(frame 1's value at XY) + (frame 2's value at XY) + (frame 3's value at XY) + (frame 4's value at XY)] / 4.

The *AstroStack* program also features additional useful image-processing algorithms, including unsharp masking, deconvolution, convolution, and histogram functions.

Tricolor Composites: Color Images from a Black-and-White Camera. Stacking images also gives you the ability to obtain color images with a black-and-white camera (Figures 8.15, 8.16). Begin by capturing successive images of your subject taken through a tricolor filter set. The most widely used tricolor filter set consists of a Wratten #25 red filter that transmits light of 600 to 700 nm, a Wratten #58 green filter that transmits light of 500 to 600 nm, and a Wratten #47 blue filter that transmits light of 400 to 500 nm. (Note: With smaller apertures, the Wratten #47 may be too dense, so a lighter Wratten #38A filter must be employed.) Such an RGB filter set mimics the spectral response of the color receptors in the human eye. Red, green, and blue are additive primary colors: red plus green gives yellow; blue plus green gives cyan; blue plus red gives magenta. It isn't intuitively obvious that yellow light is produced by mixing red and green, but this can be demonstrated by superimposing red and green spotlights on a white screen.

With rare exceptions, the CCD array in a black-and-white video camera is quite sensitive to light in the near-infrared part of the spectrum at wavelengths from 700 nm out to as much as 1,100 nm in many cameras. The filters in most tricolor filter sets "leak" in the near-infrared, so you must use an infrared-rejection filter in conjunction with them or your color composites will be distorted. (Some black-and-white cameras are equipped with removable infrared-rejection filters; all single-chip and three-chip color cameras contain integral infrared-rejection filters.)

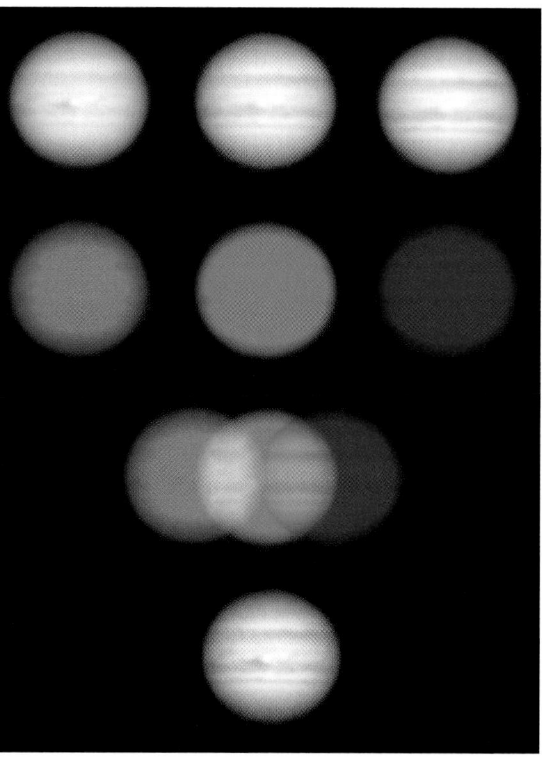

FIGURE 8.15

The sequence of steps in creating a tricolor composite image of Jupiter. At top we see each monochrome 8-bit raw image as it was recorded through red, green and blue filters (left to right). In the middle rows, each image is applied to its corresponding color channel in Adobe Photoshop, then carefully superimposed to produce the final 24-bit tricolor image at bottom.

Bear in mind that the CCD chips in the overwhelming majority of black-and-white video cameras are considerably more sensitive to red light than to green or blue light. Compared to the red frames, the green and especially the blue frames will appear grainier due to their poorer signal-to-noise ratios. To produce realistic and accurate color images you'll need to properly weight the various monochrome frames when you add them together to create a composite image.

Recently, an increasing number of astroimagers have replaced their RGB filter sets with a CMY set composed of cyan, magenta, and yellow filters. Cyan, magenta, and yellow are the subtractive primary colors. Each of these *dichroic* (bicolored) filters transmits a larger portion of the spectrum than the individual filters of an RGB filter set. The cyan (minus red) filter, for example, transmits the wavelengths passed by both the blue and the green filters of the RGB set. The magenta (minus green) filter transmits both blue and red light, while the yellow (minus blue) filter transmits both green and red. By passing more light, these filters shorten exposure times with still cameras and improve the signal-to-noise ratios of

FIGURE 8.16

This color image of Jupiter was assembled from monochrome images obtained with a video camera containing an extremely light-sensitive Sony HAD CCD array and a tricolor filter set from True Technology Ltd. Note the delicate robin's-egg blue hue of the numerous festoons in the planet's Equatorial Zone.

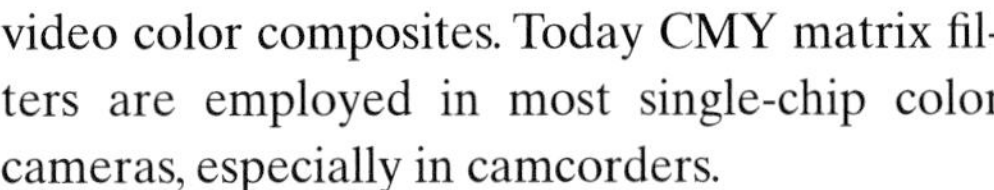

video color composites. Today CMY matrix filters are employed in most single-chip color cameras, especially in camcorders.

Fumbling in the dark to change filters and reorient the camera in order to make RGB or CMY tricolor composite images can be exasperating. To facilitate the rapid switching of color filters, a variety of filter wheels are available commercially (Figure 8.17). These devices are placed between the telescope's focuser and the video camera or projection optics. By rotating the wheel, filters can be selected with the flick of a finger, aided by click-stop detents at each filter position. Elegant motorized versions are offered by a number of firms.

From PC to Video

At some point you may want to transfer some of your images or AVI movies onto videotape for presentations. Some frame-grabber cards come with video output connectors specifically for this purpose. There are also scan converters available that connect a PC's SVGA output connector to a VCR's input jack.

FIGURE 8.17

A filter wheel is a virtually indispensible accessory for making tricolor composites. The elegant motorized unit shown here is attached to a flip-mirror device.

Storing Your Images and Movies

Computer hard drives have greatly decreased in physical size while increasing in storage capacity. However, like any magnetic medium, they are vulnerable to crashes and viruses. Bearing this in mind, you may find over time that your movie files and stills are taking up an excessive amount of disk space. To free up this valuable "real estate" and better preserve your work, we recommend that you transfer your files to some external storage media. These days there is a variety of options available, depending on your particular storage requirements. These include the common CD-R/W, DVD-R/W, large-capacity removable hard-disk drives, portable Zip or Rev drives by Iomega, or convenient USB 2.0 high-speed flash drives. Writeable and rewriteable CD-ROM and DVD drives are quite economical today, and while a CD-R can store around 700 megabytes, a basic DVD-R can store nearly seven times as much data.

Sequential Movie Player

Once you've captured several AVI movie files, you may want to view them as a continuous sequence. Unfortunately, the standard Microsoft Media players only allow you to open one file at a time. This can be quite frustrating when you have 30 or more sequences to review. If you enjoy computer programming and are familiar with Microsoft Visual Basic, compile the code in Appendix I of this book, where you will find a simple and useful sequential movie player that will make the task of reviewing your recordings to select the most promising sequences both relaxing and enjoyable. Just select the first movie in the sequence, sit back, and enjoy the show!

The Internet

There are literally thousands of astronomy-related Internet sites maintained by NASA and other government agencies, research observatories, planetariums, universities, professional and amateur astronomical organizations, and the home pages of hundreds of amateur astronomers. The wealth of information can be overwhelming. Millions of people around the world were able to view the Pathfinder pictures from the surface of Mars almost as quickly as they were downloaded and processed by NASA. Pictures of the 1998 and 1999 Leonid meteors from the best vantage points around the world were displayed at various Web sites within hours of the shower's maxima. Many who were unfavorably situated or clouded out were nonetheless able to witness the 2004 transit of Venus virtually in real time through live webcasts.

One of the major benefits of digitized images is that they make it possible for us to share the results of our hobby with colleagues around the world quickly and easily via e-mail. A Web page is also a great way to share your images with other interested amateurs. Designing a home page has never been easier with the large variety of HTML-editing programs available. Even the *Microsoft Office* package allows you to save documents in HTML format suitable for viewing with a Web browser. Further, the modest pixel resolutions offered by basic video cameras produce impressive images on standard monitors with resolutions up to 100 dpi. Images displayed on Web pages are usually GIF- or JPEG-format compressions, which facilitate quick download times for Web browsers. Animations from video images can also be converted from large AVI formats to compressed MPEG or animated GIF formats.

Hard Copy

9

Watching a videotape delivers a powerful sense of "being there" that no still image can evoke. Inevitably, however, you will have the desire to produce hard-copy versions of scenes from your videotapes. You may want to make prints of your images to frame and hang on the wall, mount in an album, or mail to friends. How can such pictures be obtained while maintaining the often superb quality of the images that you've worked so hard to record on videotape? In this chapter we'll explore three alternatives: photographing the screen of the video monitor, printing digitized images from a computer with a desktop printer, and using a video printer.

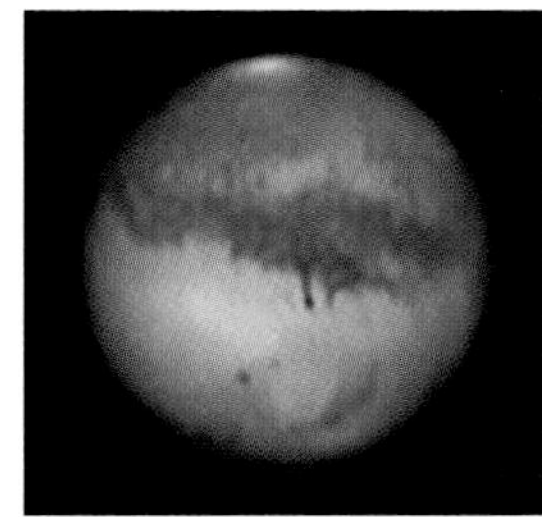

FIGURE 9.1

These images of Mars and Saturn show the ability of video cameras to capture good planetary views. This tricolor video image of Mars was captured on August 28, 2003, with an Astrovid 2000 camera during the planet's closest approach to Earth. Saturn was imaged using a tiny, inexpensive (under $90) black-and-white surveillance camera outfitted with red, green, and blue filters. Note the subtle inner C or Crepe Ring. This image is somewhat grainy due to effects from the camera's automatic gain control. When AGC affects a video sequence, several individual images can be stacked to reduce background noise and produce a more pleasing result.

Photographing the Monitor

You might imagine that simply employing the VCR's Freeze Frame mode and taking a photograph of the screen of the monitor will quickly and easily extract a nice picture from a videotape (Figure 9.1). Any attempt to do this, however, will only teach harsh lessons about the effects of electronic noise, atmospheric turbulence, and the physiology of vision.

From previous chapters we know that individual video fields are 1⁄50 or 1⁄60 second exposures and that two fields (odd and even) are interlaced to produce a single video frame displayed at a rate of 25 or 30 frames per second when the videotape is played. In the Freeze Frame mode, the single frame displayed by the VCR will invariably exhibit a poor signal-to-noise ratio that gives it the grainy, salt-and-pepper look of an overenlarged photograph. Moreover, even when seeing conditions are favorable, most frames exhibit the distorting effects of atmospheric turbulence despite these very short exposure times. Small, erratic displacements of successive images around a mean position may also be evident.

Like photographic film, the combination of your eye and brain is to some extent an integrating rather than an instantaneous sensor, requiring 1/15 to 1/5 of a second to perceive an image. When the images in a movie or videotape are viewed at rates above the threshold for flicker fusion, all you perceive is a sharp, vivid image.

To achieve the best results when photographing the monitor, set your camera's shutter speed within a range of 1/8 to 1/4 second. These exposure times will effectively combine 8 to 16 video fields (the video system's discrete 1/50- or 1/60- second exposures) to produce an appealing composite image. The improvement in the signal-to-noise ratio will be proportional to the square root of the number of frames combined, so the granularity due to electronic noise will be reduced by a factor of 2.8 (the square root of 8) to 4 (the square root of 16) compared to any single frame. Recall from Chapter 8 that the process of combining frames to reduce noise is called *recursive filtering*.

FIGURE 9.2

When photographing the video monitor, it is advisable to support the 35-mm SLR camera on a tripod for stability. Note the scan line on the monitor captured by the 1/60-second shutter speed of the camera that took the photograph.

To photograph the screen, begin by mounting your 35-mm camera on a sturdy tripod placed in front of the screen (Figure 9.2). Next make sure that the image of the entire screen fills the camera's viewfinder. Zoom lenses are far more convenient than lenses of fixed focal length, which require moving the entire camera and tripod to frame the picture. You'll want to use a cable release to avoid jarring the camera when you trip the shutter. If you don't have a cable release, use the camera's built-in self timer.

Draw the curtains to ensure that any ambient light will not scatter within the phosphor of the screen, reducing contrast. Lights can also reflect off the monitor's faceplate and annoyingly escape detection until the film is processed, so be sure to switch off any lamps or other light sources.

Adjust the monitor's brightness and contrast controls until the image is pleasing to your dark-adapted eye. Its best to adjust the monitor's contrast to ever-so-slightly less than what your eye perceives as the optimum setting.

When photographing videotapes of the planets, a camera with spot metering will precisely determine your aperture setting, but those with average metering invariably detect the surrounding dark background and overexpose the disk of the planet. Under no circumstances should you reduce the exposure time to less than $\frac{1}{8}$ second, or dark horizontal bands will appear on the photograph, recording the areas of the screen that were exposed to fewer scans of the TV tube's electron gun.

With a shutter speed of $\frac{1}{8}$ second, an aperture setting of f/5.6 to f/8 will give good results with black-and-white ISO 400 film like Kodak's Tri-X or T-Max 400. The use of slow, fine-grained films will not provide any advantage because resolution will be limited by the resolution of the recording and the monitor, which won't exceed 400 lines with S-VHS or Hi8 recordings.

If you use color print or slide film to photograph a black-and-white image, you'll probably find that your pictures have a cold, bluish cast. Most color films are balanced for the spectral properties of sunlight and exhibit this color shift under fluorescent lights or when photographing a TV or computer monitor. Shooting with a Wratten #85A light salmon color-compensating filter over the lens of your camera will eliminate this effect.

The key to success is to review the videotape carefully, noting the brief intervals of the best seeing. This is an ideal way to spend cloudy nights! Remember the eye-brain combination mimics a shutter speed of $\frac{1}{15}$ to $\frac{1}{5}$ second, so a $\frac{1}{8}$-second exposure of the running tape will be capable of capturing an image that's indistinguishable from that perceived by the eye, providing that you manage to trip the shutter at the right instant. There is admittedly some element of hit-or-miss involved, but with practice success is all but guaranteed if you have taken four or five photographs in the hope that one will turn out to be sharp.

Even under the best seeing conditions, it's rare for any two consecutive video fields or frames to be equally sharp. Unfortunately, you won't be compositing non-consecutive frames when you photograph the screen. Consequently, your pictures won't be as sharp as the sharpest individual frames, but they'll be a lot less noisy.

Bear in mind that by just using this simple technique you will have an insuperable advantage over a conventional photographer using a film camera at the telescope when it comes to capturing fine details on the Sun, Moon, and planets. The photographic exposure required to record details in an enlarged image of Jupiter using a fine-grained film like Kodak's Technical Pan 2415 ranges from 1.5 to 3 seconds, while more distant and dimmer Saturn requires exposures of 4 to 8 seconds. Even at the world's finest observing sites, perfect images seldom last for more than a second or two. That's why professional astronomers are investing millions to install adaptive-optics systems on their telescopes and why taxpayers have spent over $2 billion to put the Hubble Space Telescope into orbit.

By photographing the monitor with an exposure time of ⅛ to ¼ second as you play back a videotape, you're minimizing the exposure time — and the opportunity for atmospheric turbulence to blur the image — by a factor of as much as 64! And unlike the conventional photographer, if you don't trip the shutter at just the right moment, you can always rewind the tape and try again!

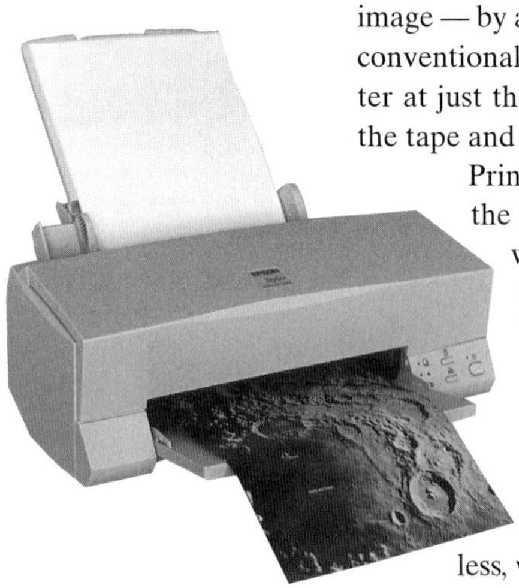

Prints and slides obtained by photographing the monitor are like any other photographs with one important exception: the degree to which they can be enlarged is limited by the resolution of the monitor, not by film grain. When stills are made from S-VHS or Hi8 recordings having 400 TV lines of resolution displayed on a high-resolution monitor, 2-by-3-inch prints appear virtually grainless, while 4-by-6-inch prints still have a very aesthetically pleasing appearance.

FIGURE 9.3

The Epson Stylus line of inkjet printers is extremely versatile and economical, capable of producing outstanding photo-quality images on special glossy papers.

Desktop Printers for Computers

Like so many consumer electronic devices, the prices of desktop printers have fallen dramatically (Figure 9.3). The combination of a moderately priced inkjet printer and coated paper will make printing your images a breeze.

The resolution of a digital printer is specified by two parameters: the number of bits of information per printing element and the number of printing elements per inch, usually expressed in dots per inch (dpi). The dots-per-inch value determines the size of the finest details that can be reproduced, while the bit value determines the subtlety of tones and hues (8-bit, 24-bit color, etc.). The appearance of the print is determined by a combination of the two. A 300-dpi, 24-bit image can look sharper than a 600-dpi, 8-bit image.

The resolution of the screen of the television in your living room is about 30 dpi. The screen of an SVGA computer monitor is typically 72 to 100 dpi. On the other hand, photo-realistic printing resolution begins at 300 dpi. So you want a printer that prints a minimum of 300 dpi. Higher is better.

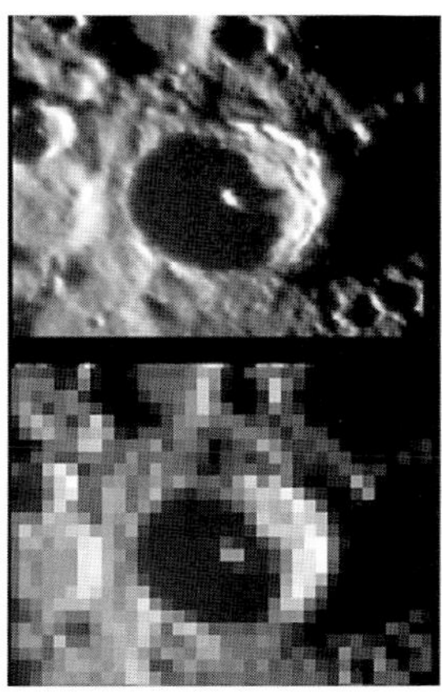

FIGURE 9.4
The image of the lunar crater Tycho (top) was captured in a window of 640 × 480 pixels, while the other image (bottom) was captured in a window of 160 × 120 pixels. Note how the larger window format provides a smoother, more realistic image when printed at the same scale.

The term photo-realistic simply means a print that looks like a photograph, with smooth edge lines. If you blow up a digital image on your computer monitor using graphic arts or image-processing software, the checkerboard pattern of its constituent pixels will soon become visible, giving edge lines a stepped look. CCD sensors have a long way to go before they can match the print resolution produced by silver halide films. A single frame of 35-mm format film contains literally billions of microscopic, light-sensitive silver bromide crystals. On the other hand, CCD sensors in the little ⅓-inch and ½-inch format chips in our video cameras are made up of a few hundred thousand pixels (the electronic equivalent of the silver bromide crystals in the film). This sets a limit on the size that we can print our video image and have it appear photo-realistic.

As a case in point, let's assume we've got a camera with a CCD array that's 450 pixels wide. A photo-realistic print will permit us to print the image at a maximum scale of 300 dots per inch. With only 450 pixels to work with, that's a mere 1.5 inches (450/300 = 1.5). Not very big!

If we were to print our image to a standard photographic print size of 6 inches, the pixels would be spread out to a density of only 75 dpi (450/6 = 75), far short of the 300 dpi required to produce a photo-realistic print. With this simple calculation we can predict that the print will appear pixelated and show unsightly jagged edges (as in Figure 9.4).

To partially overcome this restriction some graphic arts and image-processing programs have Resample or Resize commands. These add new pixels by interpolating the values of adjacent existing pixels. This artificial reproduction of the image increases its overall size while maintaining smooth edge transitions. However, it also sacrifices definition proportionally. So there is a limit to the sizes resampled enlargements can be before the image starts to look diffuse. Depending on the clarity of the image, a 25 to 50 percent increase usually works, and in some instances you may be able to get away with even more.

You may wonder why the image looks so good on a computer monitor. As mentioned previously, computer monitors have resolutions that range from 72 to 100 dpi. Hence, 450 pixels from our camera displayed on a 100-dpi monitor will produce a photo-realistic image 4.5 inches across.

If you are considering submitting your work for publication to a magazine, resolutions lower than 300 dpi may be acceptable, since magazines are generally printed at between 100 and 200 dpi. It's a good idea to contact the magazine directly about their submission requirements for electronic images.

Today's desktop inkjet printers are inexpensive and extremely versatile. Even modestly priced models are capable of producing pleasing color images at 360 dpi. In conjunction with high-grade glossy papers, they can produce images of astounding photographic quality.

They achieve this level of performance by spraying up to a million microscopic droplets of ink per second onto the surface of the page. These inks are capable of producing brilliant, saturated colors that display well. Very tiny dots of the primary colors are mixed in random patterns (*dithering*) to simulate continuous-tone printing, relying on the limitations of human vision to blur the resulting matrix enough for the dots to merge. All of the colored dots are the same size, but by carefully controlling the number of dots of each color in a given area an amazing array of hues can be replicated.

Practical ink pigments do not produce perfect primary colors, so superimposing all three primary colors at full strength creates muddy browns rather than a good black. Consequently, modern color printers use four inks: cyan, magenta, yellow, and black. This combination is known as CMYK color (the K stands for blacK). The newer, more sophisticated (and expensive) photo-type printers add another pair of inks to the mix (typically a light cyan and a light magenta) for a total of six colors.

Manufacturers like Canon, Epson, and Lexmark include software with their printers to enable users to adjust the color balance and levels of contrast and brightness of a print without affecting the corresponding values in an image file. Many people find that without making these adjustments, their prints lack contrast and are too dark. The reason is simple: it's natural to crank up the brightness and contrast levels on a computer monitor, especially if you work in a well-lit environment with windows that admit sunlight.

Many inkjet printers also include advanced self-maintenance features for cleaning the ink jets and aligning the print heads. Use these functions routinely.

Reasonable print images can be achieved using the most inexpensive grades of printer or photocopier paper, but for the best there's just no substitute for the glossy paper they use. Inkjet printers give the best results when they spray ink onto a nonporous surface that keeps the ink from diffusing or bleeding. To ensure a beautiful finish, it is also important to select the correct settings from your printer software dialogue box for your particular combination of paper and print quality.

Video Printers

If you're averse to using a computer, you'll need a video printer if you want to make photographic-quality prints of your video images at the push of a button. Video printers are seldom employed today because they fetch prices that can purchase the far more versatile combination of a computer, a frame grabber, and an inkjet printer. In addition, compositing video frames to produce images with superior signal-to-noise ratios is precluded.

A video printer grabs frames and converts them into a digital image using an integral A/D converter. Each pixel is assigned numerical position and intensity values. The brightness and contrast of the digitized image can be adjusted before printing.

Older black-and-white video printers employed a spinoff from the xerography process used in photostatic copy machines. An electrostatically charged drum transferred powdered toner onto a paper substrate where it was fused in place by applying heat. These devices were supplanted by thermal-head printers that converted different brightness values in the signal into varying degrees of heat that replicated the corresponding levels of gray on rolls of special paper. Thermal-head printers are still widely employed in medical-imaging applications, but since their printouts fade quickly over time, they are gradually being replaced by inkjet printers.

The more affordable color video inkjet printers combine ink from four ribbons to produce full-color images. The top of the line units employ a technology called *dye sublimation* printing. A solid sheet of colored ink is heated until it vaporizes and is then deposited onto the paper. Full-color images are created in four passes. The resulting prints have continuous, smooth color transitions rather than the dithered color of closely bunched dots.

Printers differ in the number of gray levels they can replicate. If that number is too small, the printed image will have an unrealistic, posterized look with abrupt transitions between an inadequate number of grays. An 8-bit printer that can produce 256 levels of gray is preferred.

Most video printers permit the user to adjust the brightness and contrast of an image before making a print. Many are equipped with a variety of other features like gamma stretching and the ability to print a negative with an inverted gray scale. The more expensive models offer many of the capabilities of a desktop computer, including the ability to adjust the size of a print, produce a multipicture printout (2, 4, or 16 pictures per page), create duplicates (6 or 16 identical pictures per page), and insert captions.

The Moon

10

The biggest, brightest object in the night sky is the Moon. It's usually the first target sighted through a beginner's telescope, yet it continues to offer challenges to even the most seasoned observers. Just as the Moon has served as the training ground for generations of astrophotographers, it will surely play the same role with future generations of astrovideographers. The skills honed imaging the Moon will prove invaluable with other subjects, especially the planets.

As the Moon waxes from a thin crescent to first quarter followed by full Moon, then wanes to last quarter and back to new Moon, observers have the opportunity to view and record its surface features under a wide variety of lighting conditions. The ever-changing interplay of light and shadow provides new vistas, even for the veteran Moonwatcher.

There is another variable that provides additional perspective to our views of the Moon, a phenomenon known as *libration.* The Moon's velocity in its elliptical orbit depends upon its distance from Earth. As Moon approaches its perigee at 356,410 km, its orbital velocity accelerates; as it recedes toward its apogee at 406,740 km, its orbital velocity slows down. All the while, however, the Moon is spinning on its axis at a uniform rate. Consequently, its rotation appears to get out of step with its orbital motion, causing an apparent wobble on its axis known as *libration in longitude* that affords a peek at a few extra degrees of longitude at the eastern and western limb regions. A corresponding libration in latitude is caused by the 6.4° inclination of the Moon's axis of rotation to the plane of its orbit, revealing extended views of the polar regions when the Moon is at opposite ends of its orbit. The combined effects of libration in longitude and latitude give Earthbound observers the opportunity to examine up to 59 percent of the lunar surface.

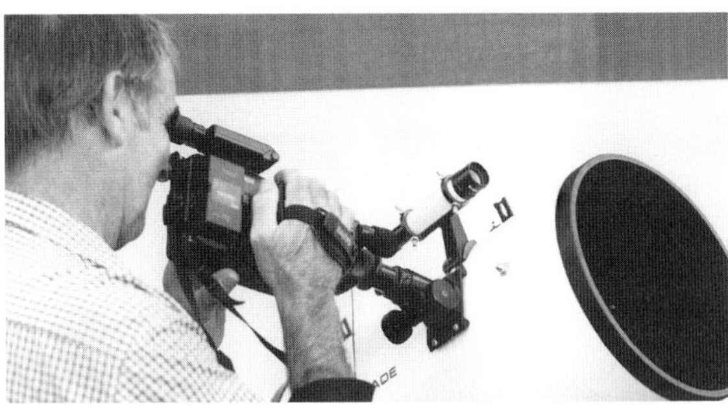

FIGURE 10.1
Pleasing images of the Moon and planets can be captured on videotape simply by holding a camcorder to the eyepiece of a telescope.

Some amateur observers may consider the Moon old hat once they've become familiar with its more prominent landmarks and showpiece formations. The truth is that many years may pass before identical observing conditions are repeated because the angles of illumination and the differences in foreshortening caused by libration are always changing. The Moon offers a virtually inexhaustible supply of subjects for your video camera!

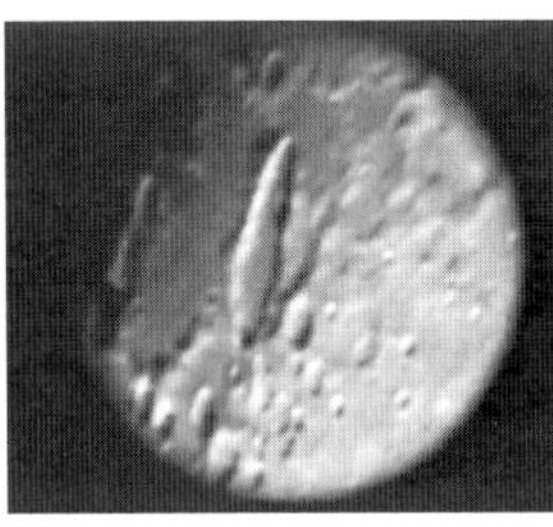

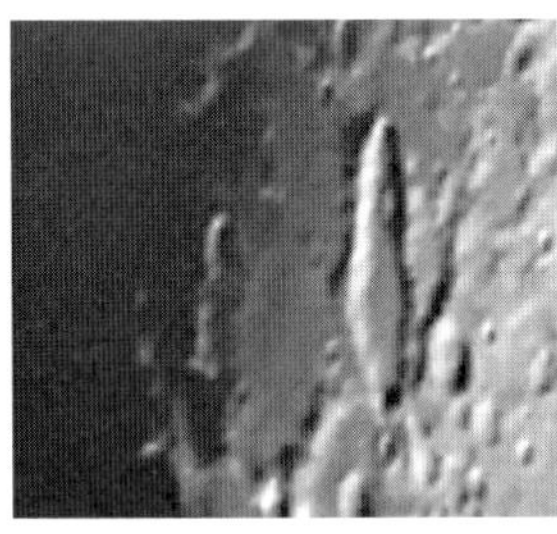

FIGURE 10.2
Left: *This image of the lunar crater Schiller was taken with a camcorder at 3× zoom held up to a 25-mm eyepiece on a 10-inch f/4.5 Newtonian. The porthole effect is due to vignetting caused by the field stop of the eyepiece.*

FIGURE 10.3
Right: *This view of Schiller was taken seconds later with the magnification of the camcorder's zoom lens increased to 6×. The field of view is narrower, but vignetting has been eliminated.*

Camcorders

The Moon is by far the most rewarding astronomical subject for anyone equipped with a camcorder. Although we unreservedly recommend purpose-built video cameras for more demanding astronomical applications, the Moon is so bright that satisfying results can be achieved using most consumer camcorders in conjunction with even small telescopes.

Few camcorders feature removable lenses, so it's likely that you'll be employing the afocal optical configuration (see Figures 1.8 and 10.1).

This simple set of instructions will help to guide you on your first attempt.

1. Set up your telescope on a night when the Moon is between a crescent phase and gibbous phase so that shadows throw topography into bold relief near the terminator. Select an eyepiece of 20- to 32-mm focal length that has generous eye relief and provides a low to medium power.

2. Make sure that the position of your telescope's focuser is easily accessible to the video camera. If you are using a Newtonian reflector, you may need to

rotate the tube so that the eyepiece can be easily aligned with the lens of the camcorder.

3. Point your telescope at the Moon. Look through the eyepiece, bring your target into focus, and center the field of view on an interesting region.

4. Set the focus of your camcorder's lens to infinity and power it up. While in the Pause mode, carefully bring the camcorder up to the eyepiece, taking pains to center the camcorder's lens directly over the eyepiece while keeping it parallel. To avoid jarring the telescope, don't let the flange of the camcorder's lens come in direct contact with the eyepiece; from an optical standpoint, a modest gap between the eyepiece and the camcorder's lens has no effect on the image.

5. You may now see a slightly out-of-focus image through the camcorder's display. Tweak the telescope's focuser to bring the lunar landscape into crisp focus.

6. At this point you may see a diffuse, dark circle surrounding (or vignetting) the illuminated lunar image. This is caused by the field stop in the barrel of your telescope's eyepiece (Figures 10.2, 10.3). Since you'll want the image of the Moon to fill the entire field of view, use the camcorder's zoom function until it does. (This is also a very convenient way to zoom in on features of interest without changing eyepieces.) Experiment with the zoom function until you achieve an acceptable tradeoff between maximum image scale and an unacceptable decrease in image brightness. The more you magnify the image, the dimmer it will become, decreasing the signal-to-noise ratio and producing an image that looks grainy. Once you find the optimum setting, you're ready to start recording.

Now that you have held a camcorder to the eyepiece, you may be sorely tempted to devise some other means of supporting it. Supporting the camcorder on a camera tripod is a very viable solution with small, short-focal-length refractors and compact catadioptric telescopes like Schmidt-Cassegrains and Maksutov-Cassegrains, especially if a star diagonal is employed. The motion of the eyepiece imparted by the telescope's motor drive relative to the stationary camcorder is not very pronounced because the distance from the polar axis to the eyepiece is so short.

Consequently, frequent repositioning of the camcorder and tripod will not be required.

FIGURE 10.4
A Canon Hi8 camcorder coupled to a 14-inch Schmidt-Cassegrain using a commercial bracket.

Larger refractors and Newtonian reflectors make the use of a separate tripod for the camcorder more problematic due to the long moment arm between the eyepiece and the axes of the telescope's mounting. If you plan to make frequent use of a camcorder with one of these telescope types, a bracket to couple the camcorder directly to the telescope (Figure 10.4) can be fashioned from wood or aluminum or purchased ready-made from one of the suppliers listed in Appendix IV.

Lunar Mosaics

The full Moon is the ideal subject for learning to assemble mosaics from video frames (Figures 10.5, 10.6). You'll need a computer with a frame-grabber card. You can capture the images for your mosaic either directly to the computer live at the telescope or from a videotape recording.

It takes eight to twelve images from a ⅓-inch-format CCD chip at prime focus of a 10-inch f/4.5 Newtonian reflector (focal length 1,140 mm) to compile a composite image of the entire face of the full Moon. This may seem an easy task, but it is actually a bit of a challenge. The most common problem is failure to precisely align the edges of the pieces of the jigsaw puzzle when assembling the mosaic, resulting in a sprinkling of ugly black squares.

1. Begin by carefully planning your capture sequence. Ensure that the telescope is well focused and the brightness and contrast settings of your software are optimized. Start at the 10 or 11 o'clock position on the Moon's disk, allowing for a bit of dark sky

around the top and left edges. At this point, locate a feature like a prominent crater or rille displayed at the bottom right on the monitor. This feature will serve as a reference point for the next image you capture. Save your sequences to disk according to the section of the Moon captured. One simple method is to designate the files by the corresponding geographic regions of the Moon, (e.g., NE.avi, N.avi, NW.avi, and so on).

2. At this small image scale, atmospheric turbulence won't be much of a problem. Capturing 1- or 2-second sequences of each section should give you an adequate number of images to find a sharp frame.

3. Move the scope horizontally to the top center of the Moon, again allowing for some dark sky at the top of your image. Check that your reference feature from step 1 is now displayed at the bottom left corner of the monitor. Capture the next 1- to 2-second sequence. Make a mental note of a feature at the bottom right of the image that will serve as your next reference.

4. Moving now to the upper right, make sure that you have dark sky on the top and to the right of the limb. Again, note that the reference feature selected in the preceding step is now displayed at the bottom left of the monitor. Capture this view.

5. At this point you must now move the telescope downward from its current position. Keep your eye on that reference crater from steps 3 and 4; it must now be positioned at the top left of our new view with some dark surrounding sky to the right. Repeat this sequence of steps, but move from right to left across the upper central region of the lunar disk. There will be no black sky background in these frames, so you will need to note features at the bottom and to the left of each view. This will ensure that you have reference features for the upper and lower central regions that will follow.

6. Once you have recorded sequences of all the required sections, sift through each sequence for the best images and save these frames as bitmap files (e.g., NE.bmp). Fire up your favorite image-editing software and create a new blank image with a black

background. You'll need a big canvas. Make sure that its pixel dimensions are larger than the combined height and width of the single frames that you'll be stitching together. For example, if you've captured 16 images, each measuring 320 × 240 pixels, and you plan to combine them into a 4-frame-by-4-frame mosaic, enlarge your canvas to 1,280 × 960 pixels. A canvas measuring 1,300 × 1,000 pixels can be cropped as required later.

7. Now open all of your images. Select one image as a reference for brightness and contrast. Move it to one side and select the frame that will overlap it in the composite image. If the brightness or contrast differ appreciably, you will need to adjust these settings with your editing software for the closest match. Repeat this process for successive overlapping frames so that the joints will not be obvious in the final composite picture. Use of a gamma-correction filter may be required for the best seamless appearance.

8. Once you have corrected for differences in brightness and contrast, select your first image at the upper left. Using your software's Edit/Copy function, copy this image to memory. Select your new canvas and again using the Edit menu, select Paste As New Selection. Move the new selection to the top left of the canvas.

9. You can now close your first image and select the second one, following the same cut-and-paste procedure as before. At this point you'll need to identify the reference crater in both images. Carefully align the second image's reference feature with the same feature in first canvas image. The other features should also line up.

10. Repeat this process for all remaining images until your masterpiece is assembled, then crop the canvas so that the Moon is nicely centered in the frame. You can now employ the enhancement functions of your image-editing software, but go easy, especially with the Unsharp Masking function! Excessive enhancement will accentuate the overlapping joints in your mosaic that you've worked so hard to make unobtrusive.

Once you've mastered the techniques of assembling a mosaic of the full Moon, a host of other rewarding lunar subjects ideally suited to this technique remain. Consider capturing a series of images of the rugged southern highlands at a gibbous phase a few days before or after full Moon. At a large image scale with the video camera oriented tangent to the lunar limb, you can create panoramic mosaics reminiscent of the views that the Apollo astronauts enjoyed from lunar orbit.

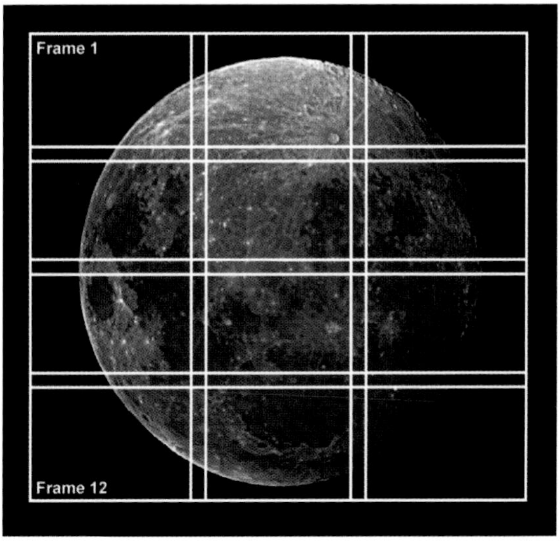

FIGURE 10.5

When making a mosaic, sections of the Moon are imaged with overlapping edges to ensure that no pieces of the "jigsaw puzzle" go missing during final assembly. The overlaps also help when adjusting the brightness and contrast settings of adjoining sections before each frame is pasted into place.

When making video recordings of the Moon at a large scale, even experienced lunar observers can lose their bearings in the lunar highlands or other densely cratered regions. This is especially true when familiar landmark formations lie outside the narrow fields of view. Changes in the angle of illumination can also be a source of confusion because the appearance of

FIGURE 10.6

Left: *This composite image of the 12-day-old Moon is made up of twelve 320 × 240 pixel images assembled like a jigsaw puzzle.*

FIGURE 10.7

Right: *No pains were taken to match the brightness of the individual frames of this hastily assembled mosaic of the eight-day-old Moon.*

many lunar formations can change dramatically under different lighting. Quickly assembled mosaics of various lunar regions under a variety of lighting conditions make excellent navigational aids for lunar videotaping sessions (Figure 10.7). Meticulous care is not required when assembling these mosaics, since the goal is simply to produce a detailed guide for personal use at the telescope. Consider labeling some features using the text function in your image-processing software.

Detailed Portraits of Lunar Formations

Smaller features can be seen on the Moon through a pair of binoculars than can be glimpsed on any of the planets through a huge telescope. The smallest features on Mars visible through the largest telescopes are about 20 km across; on Jupiter they're the size of Australia, while on Saturn they're as big as Africa. But under ideal atmospheric conditions an amateur's humble 6-inch telescope is capable of resolving craters on the Moon slightly more than a kilometer in diameter and linear features like rilles only a few hundred meters wide!

At one time or another we're all tempted to push a telescope to its theoretical limits of resolving power, which in theory is strictly proportional to its aperture, assuming the scope possesses good optical quality. The often-quoted Dawes limit states that the resolving power in arcseconds of a telescope can be calculated by dividing 4.56 by the telescope's aperture expressed in inches. This empirical formula was determined after extensive visual observations of pairs of double stars of similar brightness by the keen-eyed British amateur astronomer William Rutter Dawes, who used a variety of modest but optically excellent refractors. However, the finest details visible in the image of an extended object like the Moon or a planet are often smaller than the Dawes limit, especially if a linear marking contrasts strongly with its surroundings, like shadow-filled lunar rilles or the divisions in the rings of Saturn.

The image formed by the objective lens or mirror of a telescope is a mosaic of overlapping diffraction patterns that is in many ways comparable to the coarse-screen halftone photographs that appear in newspapers. When viewed from a distance (analogous to low magnification), outlines appear sharp and continuous, but at very close range (analogous to high magnification), the image begins to break down into a series of dots. After a certain point no additional details are revealed by further increases in magnification. When a telescope is used visually, powers exceeding 50× to 60× per inch of aperture are regarded as *empty magnification*. More often than not, long before such magnifications are achieved, atmospheric turbulence gets the final say, and the observer is forced to switch to a less powerful eyepiece.

As the aperture of a telescope is increased, the diffraction patterns in the images it produces diminish in size, analogous to decreasing the size of the dots that make up a halftone newspaper photograph. Just as there is a practical upper limit to useful magnification, there is also a lower limit below which all of the details present in a telescopic image cannot be resolved. Although this value varies from observer to observer due to differences in visual acuity, it usually falls between 13× and 20× per inch of aperture.

Recording the finest details in a telescopic image with a video camera also requires a minimum image scale, which is determined by a tenet of information theory called the *Nyquist Sampling Theorem.* This criterion states that if resolution losses are to be avoided, samples must be no larger than half the size of the finest details in the image. With the CCD array in a video camera, at least two pixels must be used to sample each resolution element in the image. The image scale in arcseconds per pixel, I, is calculated using the formula

$$I = (206)P/F,$$

where P is the pixel size in microns and F is the effective focal length of the telescope expressed in millimeters.

Let's take the case of a typical video camera containing a CCD array composed of 10-micron pixels placed at the focus of a telescope of 8 inches (200 mm) aperture with a focal length of 2,000 mm (values representing one of the popular Schmidt-Cassegrain telescopes like the ubiquitous Celestron C8). In this instance, the image scale in arcseconds per pixel is equal to (206)10/2,000, or 1.03 arcseconds per pixel. However, according to the Dawes formula, the resolving power of an 8-inch telescope is 4.56/8 = 0.57 arcseconds. According to the Nyquist criterion, the image is severely undersampled, and the details recorded will be no larger than those contained in the image produced by a telescope of only (1.03/2)4.56 = 2.35 inches aperture!

In order to satisfy Nyquist, it is necessary to increase the focal length of the telescope so that the image scale in arcseconds per pixel is equal to 0.57/2 or 0.285. If we substitute this value into our formula and calculate the required value for F, we arrive at an effective focal length of 7,017 mm. This effective focal length corresponds to a focal ratio of f/35 (7,017/200).

These rather laborious calculations can be avoided by following a simple rule of thumb. Multiply the pixel dimension expressed in microns of your camera's CCD array by 3.5. This number is the minimum focal ratio required to satisfy Nyquist. For example, if the CCD array in your video camera has 8-micron pixels, a focal ratio of f/28 will be required to record the finest details present in the image. Increasing the effective focal ratio beyond this point would only result in losses in image brightness and a deterioration of the signal-to-noise ratio.

Increasing a telescope's effective focal length can be achieved by projecting an enlarged image using an eyepiece or a Barlow lens. If eyepiece projection is employed, the amplification A of the focal length of the telescope is given by the formula

$$A = (D-f)/f,$$

where D is the distance between the video camera's CCD array and the center of the lens elements in the eyepiece; f is the focal length of the eyepiece. For example, if a 20-mm focal length eyepiece is employed at a distance of 100 mm from the CCD array, the amplification factor will be (100–20)/20 or 4×. In the case of the 8-inch Schmidt-Cassegrain telescope in our example, the effective focal length would be increased from 2,000 mm to 8,000 mm, satisfying the Nyquist criterion.

If a Barlow lens is employed, the amplification factor A is given by

$$(D+f)/f,$$

where D is once again the distance between the center of the lens element and the CCD array of the video camera and f is the focal length of the Barlow lens. With rare exceptions Barlow lenses are sold with designations like 2× or 3×, and the focal length of the lens is not provided by the manufacturer. However, it can be easily measured by performing the following procedure: on a piece of stiff white paper, draw a circle with a diameter twice the clear aperture of the Barlow lens. Allow parallel light from the Sun to fall axially on the lens and intercept the diverging cone of light transmitted by the lens with the paper. Vary the distance between the lens and the card until the circle of light exactly fills the circle; this distance corresponds to the focal length of the lens.

Most discriminating astrovideographers prefer to project an enlarged image using a Barlow lens rather than an eyepiece because Barlows usually have fewer optical elements, resulting in higher light throughput and greater freedom from internal reflections. A pair of Barlows can be stacked to provide the large amplification factors required by telescopes with fast focal ratios.

The large image scales and long effective focal ratios required to satisfy the Nyquist criterion are the reason that video cameras with excellent light sensitivity (i.e., low lux ratings) are required for high-resolution lunar and planetary imaging. As we noted in Chapter 2, black-and-white cameras are markedly superior to color cameras in this respect. Fortunately, the Moon is essentially a gray target, with only very subtle tints due to localized variations in soil chemistry, so there is little if any sacrifice suffered when a black-and-white camera is employed for lunar imaging. A yellow (Wratten #12 or #15), orange (Wratten #21), or red (Wratten #23A or #25) filter will cut through any haze that might be present and often improves seeing by a point or two on the 0 to 10 scale.

The fast frame rates of video can't always overcome the rippling of the image, especially at the large image scales required to capture the finest details that even a 6- or 8-inch telescope is capable of revealing. But even on nights that are far from ideal, there are usually a few fleeting moments of tranquil air that permit incredible views of rilles, scarps, peaks, and the terracing of crater walls to be recorded. We have included a gallery of video images of some showpiece lunar formations to demonstrate the remarkable results that can be achieved (Figures 10.8–10.22).

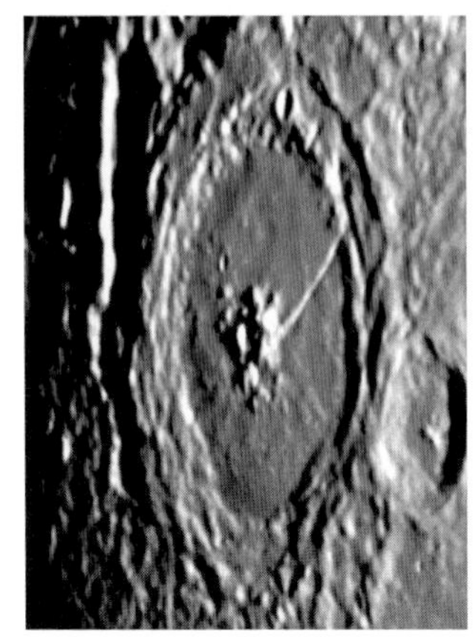

FIGURE 10.8
Although the 177-km-wide crater Petavius is actually nearly circular, it lies sufficiently close to the lunar limb that the effect of foreshortening makes it appear as an ellipse. A magnificent cleft connects the crater's massive complex of central peaks to its rim.

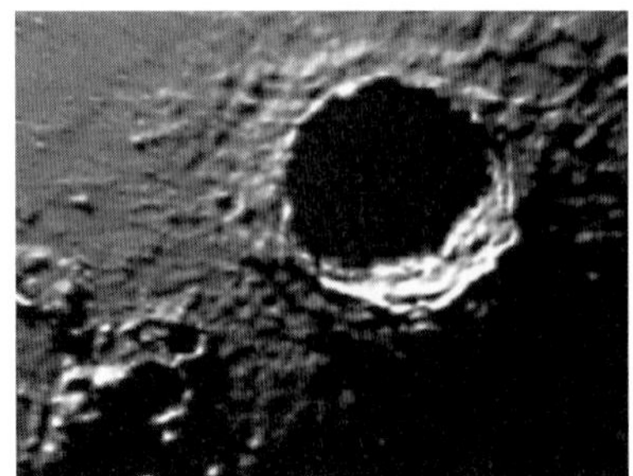

FIGURE 10.9
One of the most dramatic lunar subjects to capture on video is sunrise over the fresh, well-preserved crater Copernicus, an event that takes several hours to unfold.

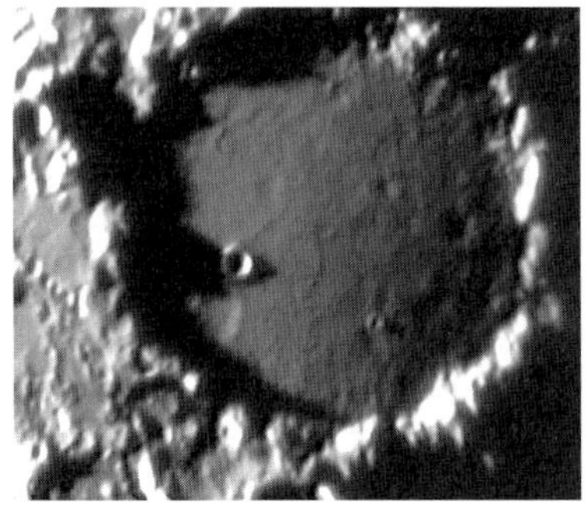

FIGURE 10.10
This image of the floor of the 153-km-wide crater Ptolemaeus shortly after sunrise captures the low-relief floor detail visible under grazing lighting.

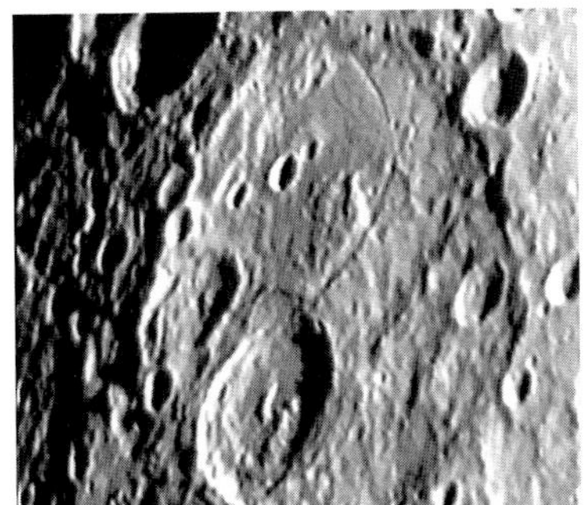

FIGURE 10.11

The ancient formation Janssen dominates this frame. The afternoon lighting highlights the varied floor detail, including a complex rille system.

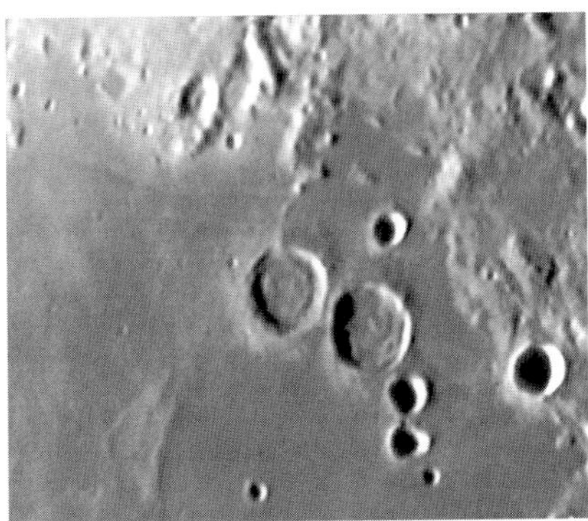

FIGURE 10.12

Morning lighting on the crater pair Sabine (left center, *30 km across) and Ritter* (right center, *31 km wide) on the shore of the Sea of Tranquillity, near the site of the first Apollo lunar landing.*

FIGURE 10.13

A larger pair of craters, Atlas (left, 87 km across) and Hercules (right, 67 km across), under late-afternoon lighting. Note the terracing of the wall of Atlas and the ruined crater Atlas E to its lower right.

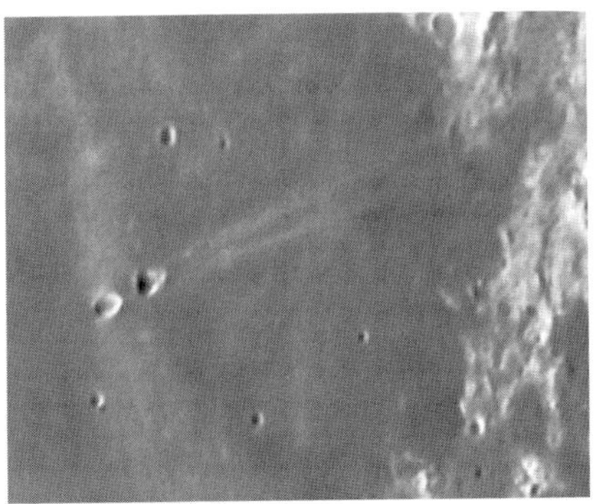

FIGURE 10.14

The crater pair Messier (left, 9 by 11 km) and Messier A (right, 11 by 13 km) are located on the Sea of Fertility. The double "comet ray" extending westward from Messier A is ejecta (impact debris) from the glancing-angle impact that formed these features.

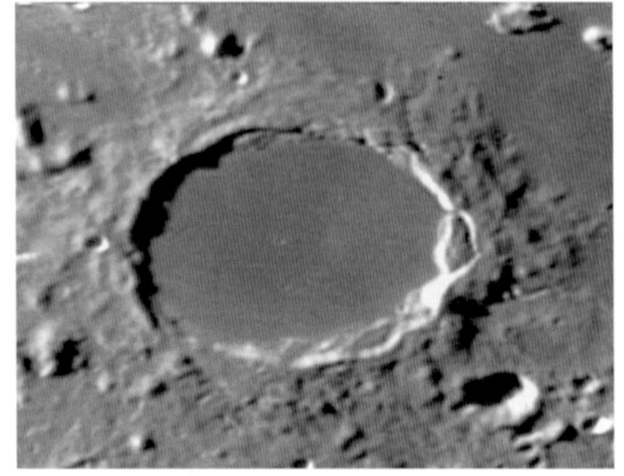

FIGURE 10.15

This image features the 101-km-wide crater Plato. The delicate craterlets sprinkled across its dark, smooth floor have served as test objects for generations of visual observers.

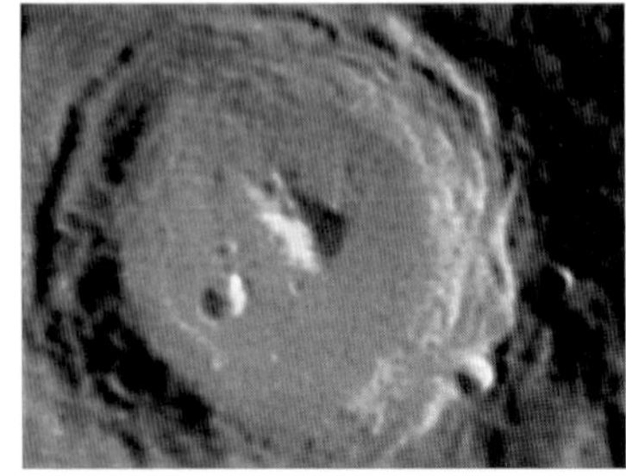

FIGURE 10.16

Measuring nearly 100 km across, the crater Arzachel has high, terraced walls and a prominent central peak.

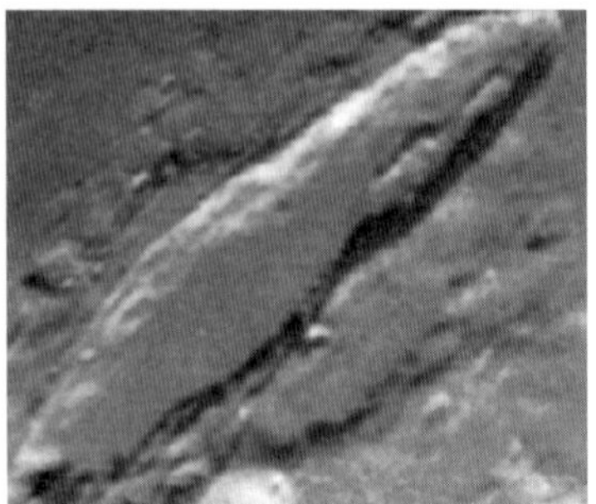

FIGURE 10.17

The highly elongated crater Schiller measures 180 km long by only 70 km wide. It is actually a compound crater, but lava flooding of its floor has effaced much of the evidence of two distinct impacts.

FIGURE 10.18

The 225-km-wide crater Clavius dominates the Moon's rugged Southern Highlands. An arc of secondary craters stretches across its convex floor.

Measuring the Dimensions of Lunar Features

Measuring the dimensions of a lunar feature using a computer (Figure 10.23) requires standard image-processing software and a decent lunar atlas such as Antonin Rükl's *Atlas of the Moon.* After selecting the formation you wish to measure, use the atlas to locate a nearby crater that has a known diameter. Next, find this reference crater in your image and measure its diameter in pixels. Now divide its diameter in kilometers by its diameter in pixels to find the distance that each pixel represents. The result is the scale of your video image in kilometers per pixel. Multiply this number by the pixel dimension of the unknown formation that you want to measure.

As an example, let's take the case of a reference crater with an established diameter of 20 km provided by the lunar atlas. Nearby is a crater roughly half its size that has no information provided. Move the

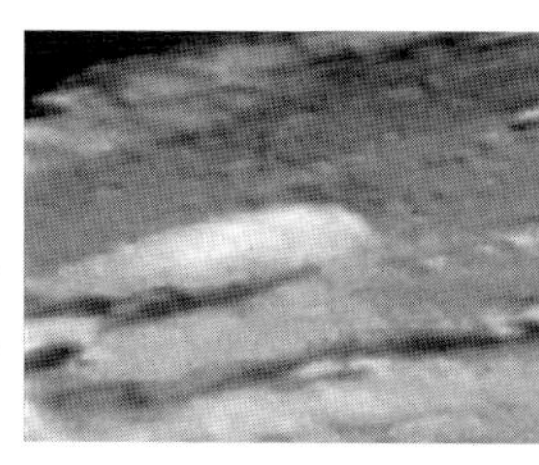

FIGURE 10.19

A portion of the ancient crater Bailly, located to the southeast of Clavius on the lunar limb. Bailly can be observed only under favorable conditions of libration in latitude.

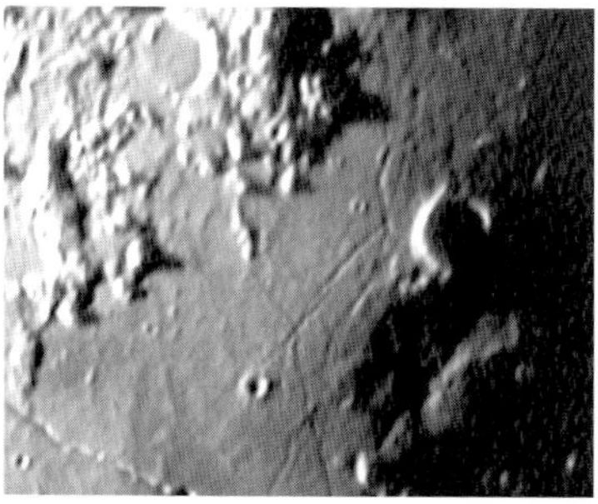

FIGURE 10.20

A low morning Sun dramatically highlights the spidery network of rilles near the 26-km-wide crater Triesnecker.

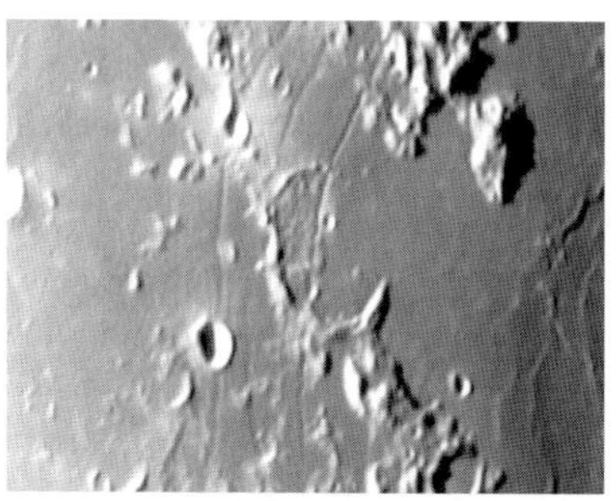

FIGURE 10.21

A curious set of three parallel rilles near the crater Hippalus arcs along the "shoreline" of the Sea of Moisture.

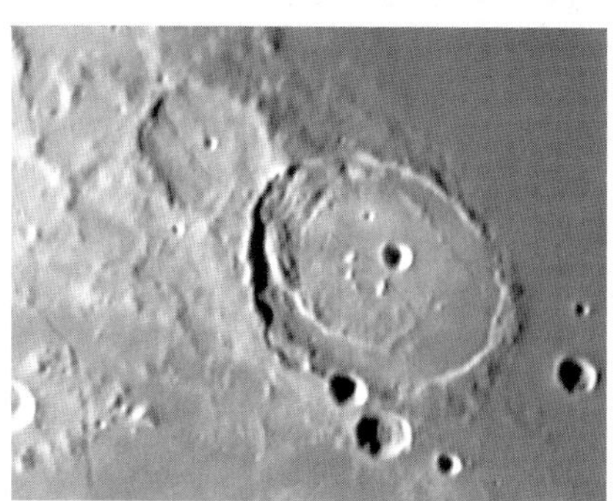

FIGURE 10.22

Almost submerged by lava flows, the floor of the 100-km-wide crater Posidonius is criss-crossed by delicate rilles.

mouse to position the cursor on the center left rim of the reference crater and note the pixel x-axis position. Let's say it reads 200.

Now, move the mouse horizontally to the right until the cursor rests on the opposite rim of the crater. Again note the current x-pixel coordinate. Let's say it reads 250. Subtracting 200 pixels from 250 pixels, we find that this crater has an apparent width of 50 pixels. Since we know from the atlas that the crater is 20 km across, we divide the two to find the distance represented by each pixel. In this example, we calculate that each pixel covers 0.4 km of lunar surface (20/50).

Turning now to the unknown crater, place the cursor on the center of its left rim. Let's say our x-pixel coordinate is 300. Moving the cursor horizontally to the opposite rim, we find our final coordinate is 330. Using the formula above, 300 subtracted from 330 gives us 30 pixels. We simply multiply 30 pixels by 0.4 km per pixel to determine the crater's diameter, in this case 12 km.

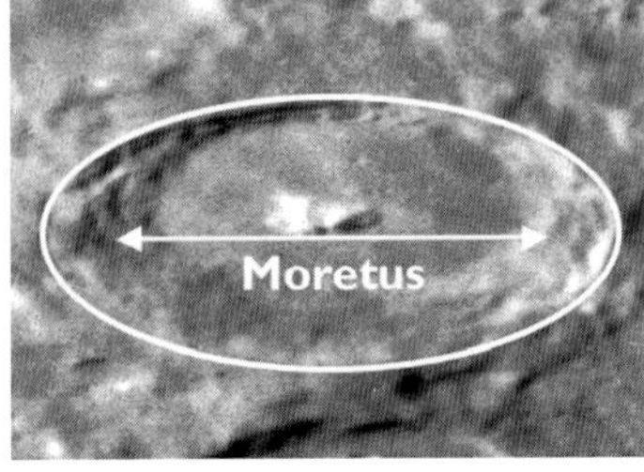

FIGURE 10.23
Measuring the dimensions of lunar features at the computer using captured video frames.

With rare exceptions lunar craters are generally circular in form, but foreshortening makes them appear noticeably more elliptical at distances of more than 40° of latitude or longitude away from the center of the Moon's disk. When using this technique, always measure along a crater's longer axis.

Transient Lunar Phenomena

For centuries, many credible observers have reported strange localized events on the Moon. Known as *transient lunar phenomena* (TLPs), they usually take the form of diffuse colored glows or obscuring hazes that can persist for several hours. Most lunar geologists are profoundly skeptical that these reports represent some sort of volcanic activity or outgassing

FIGURE 10.24
Audouin Dollfus with his video polarimeter mounted at the focus of the 1-meter Cassegrain reflector at the Meudon Observatory near Paris.

and generally write them off as misinterpretations of instrumental or atmospheric effects.

In the mid-1960s, a team of observers at the Corralitos Observatory in New Mexico monitored the Moon for thousands of hours using a then state-of-the-art image-orthicon tube TV camera at the focus of a 24-inch Cassegrain reflector, one of the first video-based professional observing programs. Even though the results were negative, today a number of dedicated amateurs continue the TLP vigil, systematically recording the lunar surface night after night in hopes of capturing one of these events on videotape. Of course, such a permanent record would be far more convincing than drawings or verbal accounts.

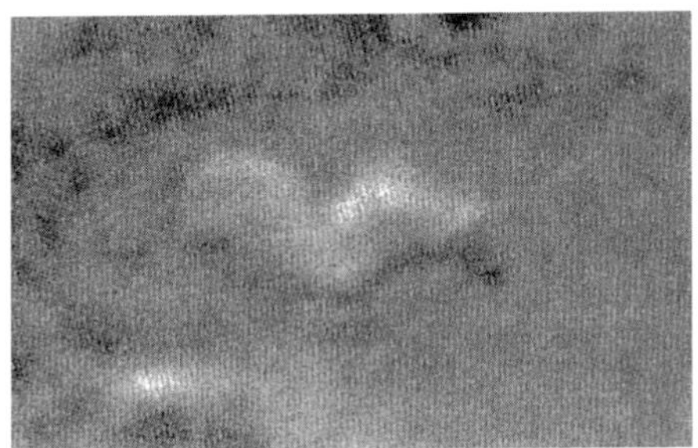

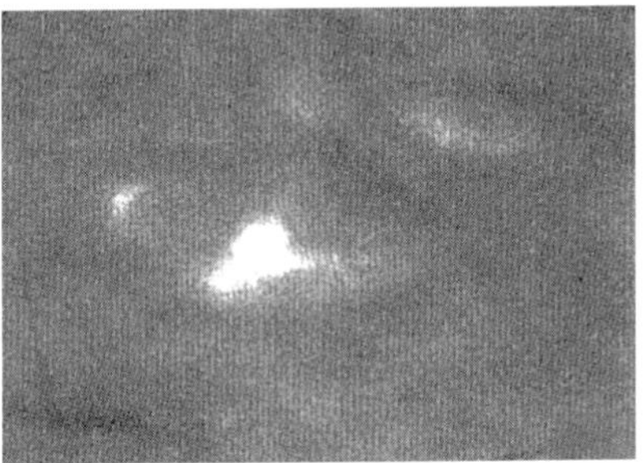

FIGURE 10.25
A pair of images through the video polarimeter showing intriguing changes in the appearance of the central peak of the crater Langrenus in 1992.

During the 1990s, the renowned French astronomer Audouin Dollfus periodically monitored the Moon using an instrument of his own design called a *video polarimeter* (Figures 10.24, 10.25). It consists of a black-and-white video camera (similar to those owned by untold thousands of amateurs) combined with a linear polarizing filter of high quality. By recording lunar formations at various orientations of the polarizing filter on successive nights and subtracting digitized video frames that represent different filter orientations, Dollfus has detected polarization anomalies on several occasions that he interprets as localized clouds of temporarily levitated dust. Although these observations are controversial, Dollfus encourages suitably equipped amateurs to pursue these investigations.

The Planets

11

If you've mastered the techniques required to record detailed video portraits of lunar craters, you're ready to test your skills on the planets (Figure 11.1). The five planets visible with the naked eye are excellent subjects, but distant Uranus, Neptune, and Pluto are so dim and remote that they make poor targets for amateur instruments.

Even mighty Jupiter at its closest approach to Earth never appears as large as a prominent lunar crater like Copernicus or Plato, and the disks of Mars and Saturn only attain about half this size. To record all the detail that your telescope is capable of resolving, you'll need to enlarge these inherently small planetary images in order to satisfy the Nyquist criterion described in Chapter 10. This demands good seeing conditions as well as patience and persistence. Capturing detailed portraits of the planets on videotape can be very challenging, but few aspects of astrovideography are more rewarding.

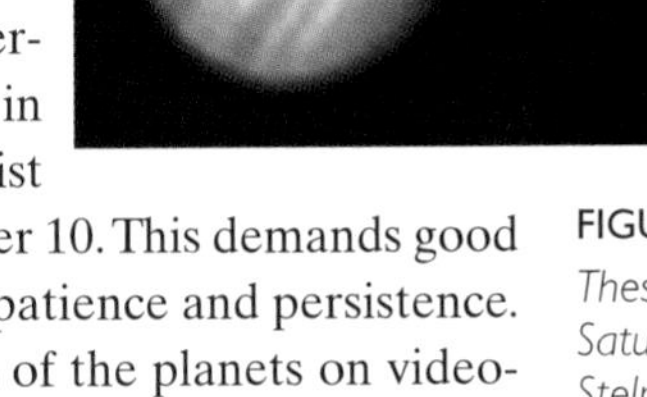

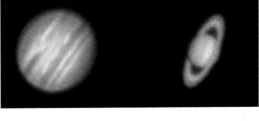

FIGURE 11.1
These pleasing images of Saturn and Jupiter by Gerald Stelmack, made with a tripod-mounted camcorder and a 5-inch f/8 Vixen Newtonian reflector, should entice other amateurs with modest equipment to record the planets.

The sections that follow are brief planet-by-planet overviews of factors of special interest to the prospective astrovideographer. Far more detailed information on the various planets can be found in the references listed in Appendix III.

Mercury

Mercury is the most neglected of the five planets that are visible with the naked eye. It is a small, airless world, not much bigger than the Moon, and never strays far from the Sun. Even during the best viewing opportunities, it must be made out low in twilight skies and observed through the densest layers of the Earth's atmosphere. Small wonder that the ancient Greeks called it *Stilbon,* meaning the Twinkling One.

When Mercury is well placed for observation, it never appears more than a paltry 10 arcseconds across. For centuries, most observers gave Mercury only a cursory glance and turned to far more rewarding targets like Mars, Jupiter, and Saturn. Mercury's markings are so delicate and easily effaced by the atmospheric turbulence that astronomers were mistaken about its axial rotation period until radar impulses were bounced off the planet with the Arecibo radio telescope in 1965.

Instead of observing Mercury low in the twilight, you'll probably have better luck if you adopt the strategy devised by Italian astronomer Giovanni Schiaparelli in the 1880s and look for the planet higher in the daytime sky. Finding the planet during the daylight hours requires a well-aligned equatorial mount equipped with setting circles or an altazimuth mount with a computerized Go To feature. Telescopic views of Mercury tend to be best early in the morning or late in the afternoon; around midday the convection produced by solar heating usually makes for "boiling" images. The presence of haze or a thin deck of cirrus clouds can obliterate the contrast between Mercury's pale disk and the surrounding sky, so observations should be attempted only on days of good transparency. Red (Wratten #25) or orange (Wratten #21) filters can be of great value in reducing the brightness of sky background.

A magnification of about 100× is required to make out Mercury's phases. You may be able to see the blunting of the southern cusp, a feature first reported by the German astronomer Johann Schröter two centuries ago; the area is one of the planet's duskier regions. If the seeing permits, a magnification of 200× will make Mercury appear about two-thirds the size of the Moon as seen with the naked eye, and you should be able to glimpse the planet's low-contrast markings. Don't be disappointed if your best efforts fail to capture anything more than the planet's phases on videotape — Mercury is a very difficult target!

By far the most detailed ground-based images of Mercury ever obtained are the work of Ron Dantowitz and Marek Kozubal from the Boston

Museum of Science, and Scott W. Teare from the University of Illinois. In August 1998 they recorded Mercury's 7-arcsecond crescent in near-infrared light using an Astrovid 2000 video camera at the Cassegrain focus of the Mount Wilson 60-inch reflector. The remarkable image in Figure 11.2 is a composite of 40 of the sharpest frames painstakingly selected from a Betacam SP-format videotape and depicts regions of the planet that were not imaged during the Mariner 10 spaceprobe's flybys in 1974 and 1975. The diffuse bright spot in the northern hemisphere almost certainly represents the blanket of ejecta (impact debris) surrounding a comparatively young impact crater. At the time the videotape recording was made the Sun was 10° above the horizon, while Mercury was at 27° — an elevation high enough to minimize the effects of atmospheric turbulence.

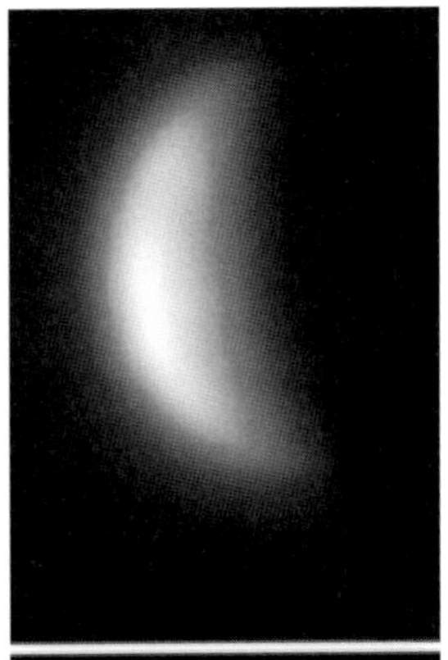

FIGURE 11.2

Top: *A composite of 40 random consecutive video fields, shows only Mercury's phase and vague hints of surface markings.*
Bottom: *Extraordinary detail appears in this composite of 40 carefully selected video fields, testimony to the power of the selective integration technique.*

Venus

Of all the planets, Venus is the most arresting to the naked eye. At its brightest it shines with an apparent magnitude of –4.4 and can be seen without optical aid in broad daylight if the sky is transparent and the observer knows exactly where to look.

Venus appears as a tiny disk only 10 arcseconds in diameter at superior conjunction, but at *dichotomy* (when the planet is 50 percent illuminated and well placed for observation in the predawn or evening sky) it subtends an apparent diameter of 25 arcseconds. At inferior conjunction it swells to a narrow crescent 65 arcseconds across.

Despite these generous dimensions, Venus is usually a blank and inscrutable disappointment to the telescopic observer. Other than during its prominent phases, which can be made out with even the smallest telescopes, it presents only an impenetrable canopy of dazzling clouds that are usually as featureless as a frosted light bulb. Distinct markings are extremely rare. Vague, amorphous mottlings and diffuse shadings can sometimes be glimpsed, but they entice few observers to a protracted study of the planet.

The phases of Venus are very easy to record by

holding a camcorder to the eyepiece of even a small telescope (see Figure 16.2). The surface brightness per unit area of the planet's cloudscape is almost 10 times greater than that of the full Moon, so you may need to employ a neutral-density or variable-density polarizing filter in conjunction with a light-sensitive black-and-white video camera, even if it has manual gain and shutter speed controls.

Within a few days of inferior conjunction, the cusps of Venus's exceedingly narrow crescent are elongated far beyond what geometry alone would suggest (Figure 11.3). The planet's dense atmosphere scatters and refracts sunlight, producing these delicate cusp extensions. They are well worth attempting to capture on videotape. Great care must be taken in sweeping up the planet when it's this near the Sun. Just as with observing Mercury during the day, red (Wratten #25) and orange (Wratten #21) filters will greatly enhance the contrast between the planet and the bright sky background.

FIGURE 11.3

The 2.9-percent-illuminated crescent of Venus near inferior conjunction. These daytime views were captured through a Wratten #25 red filter with a 12-inch Schmidt-Cassegrain operating at f/30. The image at left is a composite of 30 random video fields, while the much sharper image at right is a composite of 30 carefully selected fields. Neither image has been subjected to any subsequent image processing.

To record markings on Venus, you'll need to image the planet in near-ultraviolet light at wavelengths of 340–380 nm, just beyond the threshold of human vision (Figure 11.4). In 1927 astronomer Frank Ross of Mount Wilson Observatory made the surprising discovery that at these wavelengths Venus displays variable dusky bands that are about as prominent as the dappled plains of the Moon are to the naked eye. Although the nature of these features is not perfectly understood, it is widely believed that sulfur dioxide is responsible. This compound, which strongly absorbs ultraviolet light, has been detected spectroscopically in the planet's corrosive atmosphere. It may be emitted by volcanoes, so monitoring these somewhat chevron-shaped cloud features may provide insights about activity on the planet's inaccessible and incredibly hostile surface.

Although the light sensitivity of the CCD arrays found in most video cameras is greatly diminished in the ultraviolet region of the spectrum, Venus is so glaringly brilliant that excellent results can be achieved with most black-and-white security and surveillance cameras. Wratten #18A or Schott UG-2 or UG-11 filters will transmit the desired wavelengths while excluding visible light. If you normally use a Wratten #1A skylight filter to seal your video camera from dust, you must remove it. While it may look perfectly transparent to the eye, it blocks ultraviolet (UV) light.

When Venus is within a month or two of inferior conjunction and appears as a narrow crescent, many credible observers have seen the unilluminated portions of the planet suffused with a faint glow that is known as the *ashen light*. Often described as being reminiscent of the earthshine on a crescent Moon, the ashen light has long been regarded with skepticism by many planetary scientists, though a growing body of evidence suggests that it may be similar to the airglow that occurs high in the Earth's atmosphere when atoms ionized by solar radiation during the day release radiant energy throughout the night. A sensitive black-and-white video camera should be capable of capturing a permanent record of this elusive and controversial phenomenon.

Video also promises to be an excellent tool for studying a phenomenon known as the Schröter effect. At certain points in its orbit, Venus can appear exactly half illuminated and half shadowed — a situation called *dichotomy*. The Schröter effect is the difference of seven to eight days between the calculated and observed dates of dichotomy. The diffuse appearance of the planet's terminator due to the refraction, scattering, and absorption of sunlight in the planet's atmosphere surely plays a role in this viewing discrepancy. By varying the gain, contrast, and shutter speed controls of a video camera, the Sunlit edge of Venus may appear to change from a perfectly straight line to slightly concave or convex. The planet's apparent phase has also been reliably reported to differ appreciably in light of various colors, another phenomenon that repays careful study.

Mars

The most Earth-like world in the solar system, Mars holds more fascination for most amateur astronomers

FIGURE 11.4A
This image obtained on March 5, 2004, with a 10-inch Newtonian reflector and an ultraviolet-filtered Astrovid 2000 camera reveals the UV markings in the cloud canopy of Venus.

FIGURE 11.4B
These near-infrared views of Venus obtained in July 2004 with a 10-inch Newtonian reflector at f/10 and a Mintron 12V1C-EX integrating video camera fitted with a 1-micron filter show hints of amorphous dark markings on the planet's night side. Venus's sunlit side was drastically overexposed to record the faint emissions from the shadowed side. Compare it with the normal (short-exposure) view of the planet's crescent in the bottom image.

than any other planet. Unfortunately, it is also a difficult planet to observe. Every 26 months Earth overtakes and passes Mars in its orbit around the Sun. During these *oppositions* Mars rises in the east as the Sun sets in the west. Because the orbit of Mars is far more elliptical than Earth's, the closest ones (*perihelic oppositions*) bring the planet much closer than others. Then Mars is separated from Earth by only 55 million km and reaches an apparent angular size of about 25 arcseconds. These comparatively rare events are spaced at 15- or 17-year intervals; the last occurred in August 2003 and the next will occur in July 2018.

At the least favorable *aphelic* (most distant) oppositions, Mars comes no closer than 99 million km, and its apparent angular size doesn't exceed 14 arcseconds. Yet even at this very modest size, it is possible to capture the seasonal waxing and waning of the planet's polar caps and to monitor clouds and dust in its tenuous atmosphere.

Red (Wratten #25 and #23A) and orange (Wratten #21) filters markedly increase the contrast between the planet's ocher deserts and the dusky albedo features that early observers of the planet mistook for seas or tracts of vegetation. Covered with rocks and coarse soil particles, these regions exhibit seasonal changes in contrast as winds alternately deposit and remove fine dust transported from the surrounding plains and basins. While they are for the most part permanent, over the years they undergo interesting long-term secular changes that subtly alter their intensities, shapes, and sizes.

A day on Mars lasts 37.5 minutes longer than a day on Earth, so Martian surface features appear to cross the planet's central meridian (the imaginary line passing between its poles, bisecting the disk) 37.5 minutes later each night. If you observe Mars at the same time on successive nights, you'll find the planet's surface features displaced by 9.5° of longitude; it takes 36 terrestrial days before a Martian feature will once again appear to cross the central meridian at the original time. Consider recording Mars at the same time on every clear night for several weeks around the date of opposition in order to capture the planet's features as they move across the disk. Once you have captured a sufficient number of images, you can assemble a dramatic time-lapse animation showing the planet's rotation.

In recent years, the Mars sections of the Association of Lunar and Planetary Observers (ALPO) and the British Astronomical Association (BAA) have encouraged amateurs to monitor Martian meteorology with the aid of color filters. The clouds of dust that are periodically swept aloft by fierce winds stand out when yellow (Wratten #12 and #15) filters are employed. Green (Wratten #56 and #58) filters reveal the presence of low-lying fogs and deposits of frost on the Martian surface, while blue (Wratten #38A) and violet (Wratten #47) filters accentuate high-altitude clouds and hazes near the poles and on the morning and evening limbs.

Mars has no oceans to store heat and complicate its weather patterns, so it is an ideal laboratory for studying climatic change. Without the moderating effects of large bodies of water, the seasonal advance and retreat of the planet's polar caps are very sensitive to changes in the amount of solar radiation that strikes them. By measuring the changing dimensions of the Martian polar caps, amateurs can collect valuable scientific data.

Video has proven to be a powerful tool for imaging the planet (Figures 11.5, 11.6, 11.7), including measurements of the waxing and waning of the Martian polar caps. Traditionally, visual observers made these measurements using a device called a *filar micrometer*. The observer sighted on a pair of fine parallel wires located at the focal plane of the eyepiece. One wire was fixed, while the other could be moved by means of a calibrated screw. As the screw was turned, the number of turns or fractions of a turn required for the wire to traverse the space between two features provided a precise measure of their apparent separation.

Using a filar micrometer requires the use of very high magnifications in order to minimize the effects of a phenomenon that occurs in the eye of the observer called *irradiation*. Caused by the spreading excitation of the retina beyond the area that is

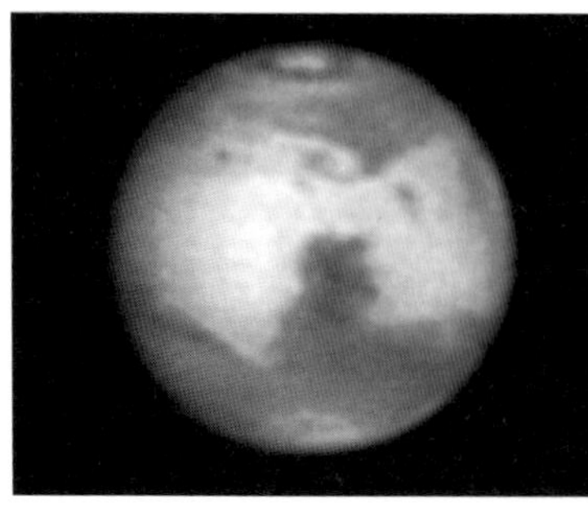

FIGURE 11.5

A monochrome image of Mars on May 1, 1999, through the 24-inch reflector at Siding Spring Observatory with an Astrovid 2000 camera. The image scale was enlarged using eyepiece projection to provide an effective focal ratio of f/36. This image is a composite of 12 video frames stacked using AstroStack software and processed with Adobe Photoshop. North is up.

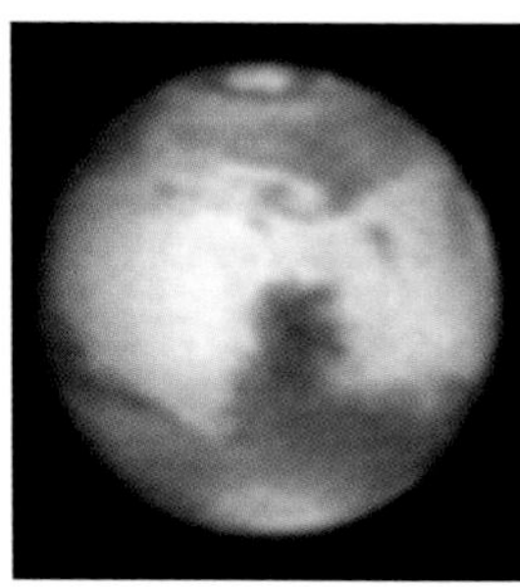

FIGURE 11.6

A low-resolution full color image obtained with a single-chip color camera has been combined with the monochrome image in Figure 11.5 to give the best of both worlds — the unsurpassed resolution of a black-and-white camera and the aesthetic appeal of color. NASA occasionally uses a similar technique to process spaceprobe images. North is up.

directly stimulated by a bright light source, irradiation impairs the accurate measurement of planetary features by making bright objects appear larger than they really are. Combating irradiation by using high magnifications places great demands on the tranquillity of the atmosphere, the accuracy of the telescope's drive, and often the observer's physical and mental stamina. Of course, conventional photography is even more at the mercy of atmospheric turbulence, since exposures several seconds long are required to record Mars at a sufficiently large image scale. In addition, the scattering of light within photographic emulsions introduces errors analogous to visual irradiation.

Videotape Mars using a sensitive black-and-white security camera or a purpose-built astronomical video camera. A red (Wratten #23A or #25) filter should be employed to penetrate the hazes that are often present and can result in spuriously large polar cap dimensions. If you digitize selected frames using a video capture card, you can make polar cap measurements using the techniques described for determining the dimensions of lunar craters in Chapter 10. Alternatively, by playing back the videotape, the latitude of the perimeter of the polar cap can be measured directly on the video monitor using an inexpensive vernier caliper. Examine the videotape frame by frame, measuring only the sharpest frames that occur during fleeting moments of stable seeing. Due to small, random displacements of the image caused by an ever-present continuum of high-frequency atmospheric turbulence and minor errors in the telescope's drive gears, you'll probably find it easier to measure discrete frames in the Freeze Frame mode rather than attempting to measure the erratically displaced image in the Play mode. Two dimensions are measured on each frame selected: the maximum east-to-west width of the polar cap (EW) and the polar diameter of the planet's disk (PD). The Martian latitude is equal to the arc-cosine of EW/PD. For maximum precision, measure 10 or 20 frames and calculate an average value.

By adjusting the monitor's brightness and contrast controls, the effects of irradiation caused by the brilliance of the polar caps can be all but eliminated. The locations of the planet's limbs can be determined with a high degree of precision if the monitor's brightness control is momentarily tweaked so that the calipers' jaws are seen in silhouette against a brightened background sky.

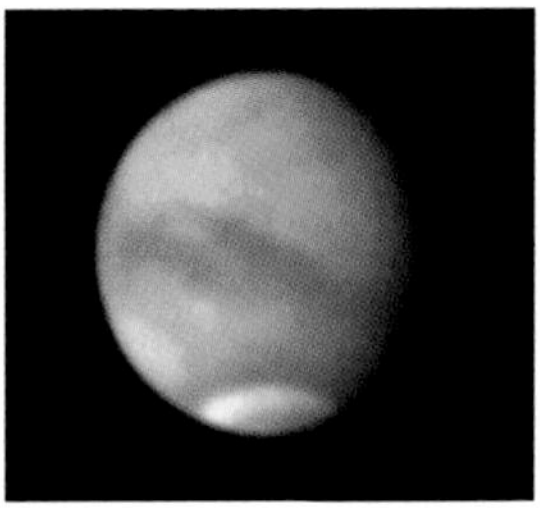

Even in the Space Age, amateur observations of Mars continue to have value to professional planetary scientists. For example, when the Pathfinder spacecraft was approaching Mars in 1997, NASA provided a Web site and FTP upload link for amateurs to submit their observations of Mars. The world wide around-the-clock record of the planet's changing weather conditions provided early warning of dust storms that might have threatened a safe landing.

FIGURE 11.7
Mars on July 16, 2003. This tricolor image was captured from Sydney, Australia, with an 8-inch Newtonian reflector operating at f/30. North is up.

Jupiter

Jupiter appears in the night sky 10 out of every 12 months, shining brightly at an average magnitude of –2. Even at its smallest apparent diameter of 30 arcseconds, Jupiter appears larger than Mars at its closest approaches to Earth; at opposition its disk can subtend an apparent diameter of almost 50 arcseconds.

Jupiter spins on its axis faster than any other planet; a Jovian day is less than 10 hours long. This furious rate of rotation is so rapid for a body of Jupiter's enormous bulk that the planet exhibits a pronounced bulge at its equator due to centrifugal force. The characteristic parallel banding of Jupiter's atmosphere is produced by this rapid rotation. The bright zones consist of wispy cirrus clouds of frozen ammonia crystals produced by the convective upwelling of heat from the planet's interior, while the intervening dusky belts correspond to regions where downdrafts prevail. At the boundaries between the belts and zones there is an endless succession of storms and eddies created as countervailing atmospheric currents slide past one another.

FIGURE 11.8

A tricolor composite image of Jupiter on October 21, 1999, at the f/18 Cassegrain focus of the 24-inch reflector at Siding Spring Observatory.

With its generous disk size, a wealth of colorful features, and a respectable apparent surface brightness per unit area (almost ⅙ the full Moon's), Jupiter is by far the most rewarding planetary target for video cameras (Figure 11.8). A camcorder held to the eyepiece will give very pleasing results, while a low-light black-and-white security camera used in combination with an 8- or 10-inch telescope can record a level of detail rivaling the best results obtained by professionals by means of conventional photography as little as a decade ago.

The contrast between Jupiter's belts and zones can be enhanced with a light blue (Wratten #80A or #82A) filter. The blue festoons that protrude into the Equatorial Zone from the southern edge of its North Equatorial Belt and many features at high latitudes darken dramatically when yellow (Wratten #12) or orange (Wratten #21) filters are employed. Yellow-green (Wratten #11) and light green (Wratten #56) filters often give the best overall views and accentuate the visibility of the Great Red Spot at those times when it exhibits a distinctly ruddy cast.

FIGURE 11.9

Jupiter seldom fails to present a host of interesting phenomena to the backyard astronomer. This sequence of tricolor composite images by David Moore shows the evolution of a rifting disturbance in the planet's North Equatorial Belt that first appeared as a brilliant white spot on November 12, 1999, then rapidly lengthened in longitude over a span of less than a week, all but disappearing by November 19th. Moore used an Astrovid 2000 camera with Barlow lens projection to achieve an effective focal ratio of f/33 on his 14-inch Cassegrain.

If you intend to capture the delicate pastel hues of Jupiter's clouds by compositing monochrome frames made through a red, green, and blue (RGB) or cyan, magenta, and yellow (CMY) tricolor filter set, be sure to obtain all the images within an interval of two minutes or less, otherwise details in the resulting color composite will be smeared by the planet's rotation.

Amateur astronomers have a long and distinguished tradition of monitoring Jupiter, ready to alert their professional colleagues of interesting developments (Figure 11.9). The principal technique employed by dedicated observers of the planet is the careful timing of the passage of the planet's ever-changing features across its central meridian. These values reveal the presence of discrete currents in the Jovian atmosphere. In one minute Jupiter turns 0.6°

on its axis, so it is quite possible to determine the longitudes of the planet's features to a precision of less than 1°, especially if timings made by several observers are pooled. While visual central meridian transit timings continue to be valuable, video has added a new level of accuracy. Video is also a boon when measuring the latitudes of Jupiter's belts and zones; the pronounced darkening at the planet's limbs that has long been a source of error when conventional photographs were measured can be easily overcome.

Jupiter's "Miniature Solar System"

Most members of Jupiter's retinue of satellites are trifling fragments of cosmic debris, but the four large satellites discovered by Galileo with his primitive telescope in 1610 are worlds in their own right. Christened Io, Europa, Ganymede, and Callisto, they range in brightness from magnitude 4.5 to 5.5 when Jupiter is at opposition and would be visible to the naked eye were they not immersed in the glare of their parent planet. Even binoculars reveal their presence. A camcorder held to the low- or medium-power eyepiece of a 4- or 6-inch telescope will record the configuration of the Galilean satellites, which changes perceptibly hour to hour as they revolve around the planet. However, it will probably be necessary to overexpose Jupiter, burning out the features on its disk (Figure 11.10).

FIGURE 11.10
Jupiter and the Galilean satellites through the 60-inch Mount Wilson reflector.

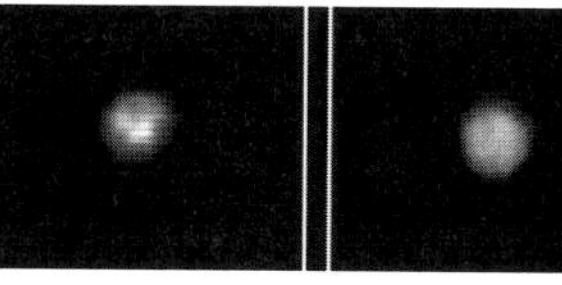

FIGURE 11.11
Hints of the delicate surface markings on Ganymede's tiny disk were recorded using eyepiece projection on a 10-inch Newtonian reflector on September 15 (left) and 29 (right), 1998.

With a diameter of 5,300 km, Ganymede (Figure 11.11) is not only the largest Galilean satellite but the largest satellite in the entire solar system — not much smaller than the planet Mercury. Markings dappling its tiny 1.8-arcsecond disk have been captured on videotape using a 10-inch aperture Newtonian, a feat beyond the grasp of all but the world's largest telescopes in the era of silver halide photography. Ganymede's dusky features have a distinctly ruddy hue. The most prominent is an oval marking called Galileo Regio, which contrasts strongly with the surrounding brighter, grooved terrain. It is interesting to note that the bright highlands on Earth's Moon are less reflective than the darkest regions on icy Ganymede.

The planes of the orbits of the Galilean satellites lie in almost exactly the same plane as Earth's orbit, so they appear to pass directly across the disk of Jupiter and then behind the planet during each circuit of Jupiter (Figure 11.12). Transits of Callisto occur much less often due to the combined effects of this satellite's orbital inclination and its distance from Jupiter. Prior to the date of an opposition of Jupiter, a satellite approaching occultation will be seen to gradually fade from view while still at a considerable distance from the planet's limb as it enters the cone of shadow cast by the planet. After opposition the shadow is located on Jupiter's opposite side, and an eclipsed satellite will reappear seemingly out of nowhere, rapidly increasing in brightness as it emerges into sunlight.

FIGURE 11.12
Diffraction-limited images of Callisto, Ganymede, and Io (upper left) captured by Ron Dantowitz through the 60-inch reflector at Mount Wilson Observatory. The halos of ejecta around Io's volcanoes appear as a diffuse dappling, (lower left).

For a period of several months at six-year intervals, the plane of Earth's orbit is virtually coincident to the plane of the orbits of the Galileans. At these times it is possible to observe occultations and eclipses of one satellite by another. During *mutual occultations,* two satellites appear to merge into one; during *mutual eclipses,* the shadow of one satellite passes across the tiny disk of the other, causing it to fade or disappear.

The shadows that the Galileans cast on Jupiter's cloud deck appear as tiny, jet black dots. Prior to the date of opposition, the shadows appear on the planet's disk before the satellites themselves; after opposition the shadows follow the satellites and remain on the disk after the passage of the satellite has ended.

These shadow transits are excellent subjects for the astrovideographer, especially for making dramatic time-lapse animations. The dates and times of these events are provided each month in the Celestial Calendar department of *Sky & Telescope* magazine.

The length of time required for the Galileans to pass across Jupiter's disk depends on their periods of revolution around Jupiter. Io, with an orbital period of 1.77 days, and Europa, with an orbital period of 3.55 days, complete full central transits in slightly over 2 and almost 3 hours, respectively. Ganymede, with an orbital period of 7.15 days takes about 3 hours and 45 minutes to cross the center of Jupiter's disk; transits of outermost Callisto, with an orbital period of 16.69 days, take up to five hours. However, shadow transits of both Ganymede and Callisto can occur quite close to Jupiter's polar regions, resulting in events of much shorter duration.

When the transiting Galilean satellites are seen against the backdrop of Jupiter's clouds, a variety of fascinating appearances can be recorded. All four appear like gleaming little pearls at ingress or egress when they are near Jupiter's dusky limb, but their appearances can vary dramatically as they pass over the face of the planet due to changing contrast with the background.

Near Jupiter's central meridian, Io appears as a very faint gray spot when superimposed on one of Jupiter's bright zones but can be seen as an elongated bright spot when it crosses one of the planet's dark belts. This distorted form is the result of Io's bright equatorial regions and dusky poles. With an albedo of 64 percent, ice-covered Europa can be very hard to detect while in transit over a bright zone. Ganymede usually appears to fade as it moves away from the limb, becomes difficult to make out for a period of about 20 minutes, and then seems to be gradually transformed into a prominent grayish brown spot. Transits of Callisto are the most spectacular of all, since the satellite rapidly changes appearance from a bright spot into an intensely dark one shortly after ingress onto Jupiter's disk is completed. Silhouetted against a bright zone, the disk of Callisto appears so dark it can sometimes be mistaken for a shadow.

FIGURE 11.13

Saturn on August 20, 1998, photographed with the Astrovid 2000 camera and a red (Wratten #25) filter at the f/16.2 Cassegrain focus of the 60-inch Mount Wilson reflector. This image is a composite of six video fields.

Saturn

Saturn is one of the most enchanting sights in the heavens. This fact is certainly not attributable to Saturn itself but to its magnificent system of rings. Saturn's globe is striped by alternating belts and zones reminiscent of Jupiter's, but their contrast is muted by the presence of an overlying layer of hazy aerosols. Moreover, Saturn's atmosphere is more quiescent than Jupiter's, so well-defined irregularities in its belts are comparatively rare; only at intervals of many years do prominent eruptive white spots appear.

Saturn's disk never appears larger than a rather paltry 21 arcseconds, but the planet is certainly within the grasp of astrovideographers (Figures 11.13, 11.14). Moreover, despite its highly reflective canopy of clouds, it is so remote from the Sun that its apparent surface brightness per unit area is less than 1⁄20 of the full Moon's. Don't be disappointed if Saturn proves to be a difficult subject.

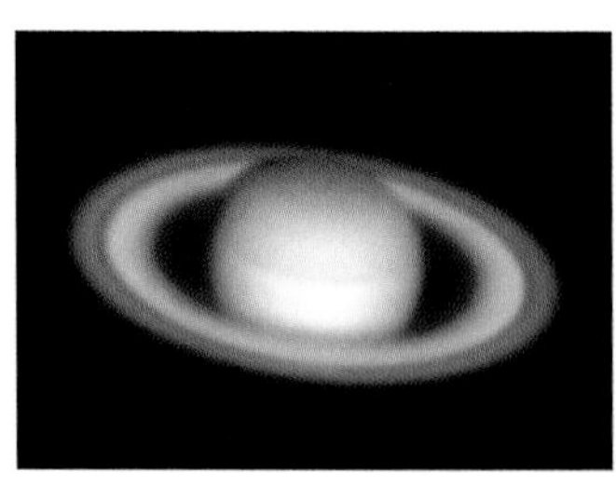

FIGURE 11.14

This tricolor composite image of Saturn was assembled from a video recording made on January 25, 2003, with an 8-inch Newtonian reflector and a simple homemade camera.

For well over a century, astronomers have designated the principal components of Saturn's ring system by the letters A, B, and C. Ring A is the outermost ring visible through amateur instruments. It has a steely blue-gray hue and is separated from the broadest and brightest ring, Ring B, by a gap 4,500 km wide called the Cassini Division. The innermost, and much fainter, Ring C is often called the Crepe Ring. Visually it is a challenging object for a 5-inch telescope, though even smaller apertures can follow where it crosses in front of Saturn's globe and appears as a narrow dusky stripe.

Every few decades, Saturn passes in front of (occults) a moderately bright star. When this last occurred in 1989, affordable low-light video cameras had not yet appeared in the marketplace, so only a handful of observers managed to record the event on videotape (Figure 11.15). These recordings proved to be both dramatic and scientifically valuable. As the 5.8-magnitude star 28 Sagittarii passed behind the rings, its brightness fluctuated abruptly. These flickerings revealed differences in the particle density of the rings far too delicate to be observed directly. As we discuss in Chapter 12, future events of this kind should not be missed by the astrovideographer!

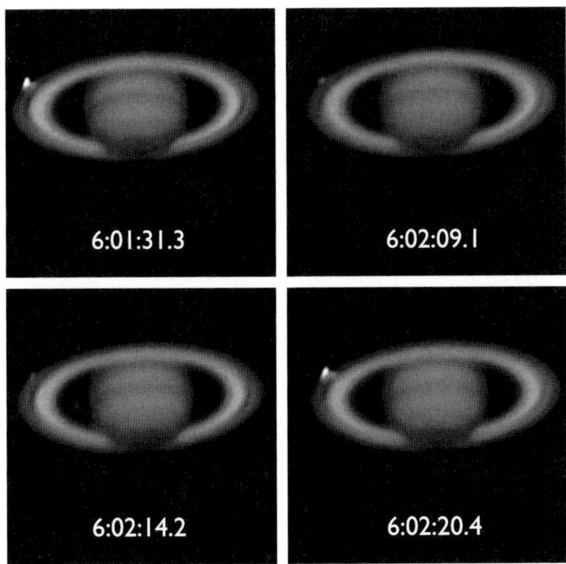

FIGURE 11.15

These images of the occultation of 28 Sagittarii by Saturn are frames from a videotape recorded by Sky & Telescope *editors Leif Robinson and Dennis di Cicco with the 60-inch Mount Wilson reflector. Scores of dramatic and rapid fluctuations in the brightness of the star appear on the videotape. Some last only a fraction of a second, while others are tens of seconds in duration. The image at top left captures the undiminished brightness of the star 37 seconds before it contacted the outer edge of Ring A. At top right, the star dims as it passes behind Ring A. As shown at bottom left, the star remained visible throughout its trek behind Ring A. At bottom right, the light of the star appears to suddenly flare as it shines through the Encke Division near the outer edge of Ring A.*

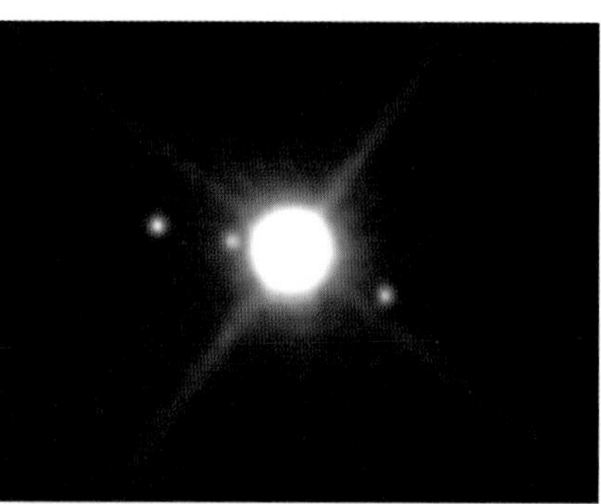

FIGURE 11.16

Recording the remote outer worlds of Uranus, Neptune, and Pluto with typical amateur instruments requires a highly sensitive video camera. Only the featureless disks of Uranus and Neptune are revealed, along with a handful of their brightest moons. The movement of tiny, distant Pluto may be recorded over several days as a dim point of light among the background stars. This view of Uranus and three of its brightest moons was captured on May 16, 2004. Although the planet was around 3 billion kilometers from Earth at the time, 14th-magnitude Titania (far left), Ariel (left), and Oberon (right) were recorded using a 10-inch Newtonian reflector fitted with a Mintron 12V1-EX video camera operating at f/15. The planet's disk was deliberately overexposed to record the satellites.

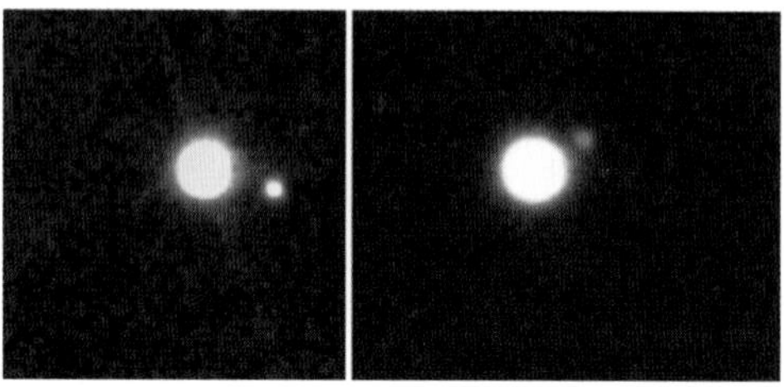

FIGURE 11.17

The same telescope and camera easily reveal Neptune and its brightest satellite, Triton. Using the Mintron camera's 96x integration mode, these images taken 22 hours apart demonstrate Triton's orbital motion. The view at left was recorded under better seeing conditions.

Occultations & Transits 12

An occultation occurs when one celestial body passes in front of a more remote one. Given the host of objects moving within the solar system — planets, satellites, asteroids, and comets — many variations of this theme are possible. Strictly speaking, a solar eclipse (which we'll get to in Chapter 13) is an occultation of the Sun by the Moon.

Lunar Occultations

So close to our vantage point on a cosmic scale, the Moon is by far the most common source of occultations because it subtends a large apparent diameter of ½° and appears to move rapidly against the starry background. As its orbital motion carries it from west to east across the sky, the airless Moon acts like a knife edge slicing through space, abruptly cutting off the light of the far more distant stars and planets. It's a fascinating spectacle. The Moon seems to slowly creep toward the tiny speck of a star. For a brief moment the star seems to hang on the very edge of the Moon's disk, then it disappears with startling suddenness. The observer who blinks an eye at the wrong instant may miss this event completely.

Time one of these phenomena accurately and you'll have a potentially valuable piece of scientific data. The sheer drama of occultations can also be compelling. Witnessing one of Jupiter's Galilean satellites emerge from behind the limb of our own satellite followed by mighty Jupiter itself is an experience you won't soon forget!

When the phase of the Moon is waxing, occulted stars are seen to disappear behind the dark Earthlit limb and reappear from behind the bright Sunlit limb (Figure 12.1). When the phase of the Moon is waning, disappearance occurs at the bright limb and reappearance at the dark limb. As the Moon's shadow sweeps across the Earth's surface, it defines a region from which the total occultation of a star is visible. North and south of this region, the star passes just off the Moon's limb.

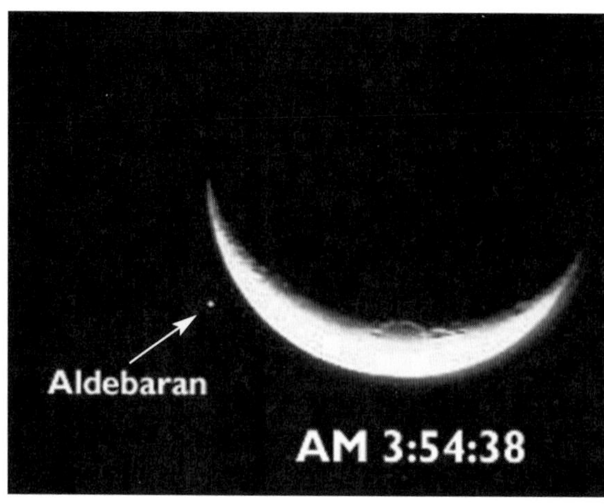

FIGURE 12.1

Even a camcorder is capable of recording occultations of planets and the brightest stars. This image of the crescent Moon and Aldebaran was recorded by Dale Ireland on April 19, 1999.

On a narrow track a little over a kilometer wide located at the northern and southern edges of the Moon's shadow, observers will see the star appear to skim along a path tangent to the Moon's northern or southern limb, repeatedly blinking on and off as it passes behind the lunar peaks and valleys. Accurate timings of the sequence of disappearances and reappearances during these grazing occultations provide very precise information about the contour and topographic relief of features at the Moon's limb. Refining our knowledge of the profile of the lunar limb aids the analysis of total solar eclipse timings, which can be used to detect minute changes in the diameter of the Sun. Such changes are suspected to occur over periods of many years and may subtly influence Earth's climate.

The data derived from precisely timing occultations are also useful for detecting minute variations in Earth's axial rotation rate. When used to measure the almost imperceptibly gradual deceleration of the Moon's orbital motion along the ecliptic caused by the tidal forces at work in the Earth-Moon system, occultation data suggest a deceleration value of 23 arcseconds per century, but theory implies a value of 28 arcseconds. Some astronomers have attributed this discrepancy to changes in the gravitational constant. If true, this would have profound implications for cosmology.

As early as 1865, the eminent British astronomer Sir John Herschel reasoned that "a double star, too close to be seen divided with any telescope, may yet be detected as double by the mode of its disappearance" when it is occulted by the Moon. Under such circumstances, a large fraction of the light of a seemingly single star will disappear abruptly, but a small portion will appear to linger for an appreciable fraction of a second. Many previously unknown or suspected binary star systems have been discovered or confirmed in this fashion, some with apparent angular separations of as little as 0.02 arcseconds. Occultations also afford an opportunity to determine the apparent diameters of some stars. In a few instances, discrete layers in stellar atmospheres have even been detected. Traditionally, high-speed photoelectric photometers were employed for this work, but such equipment is well beyond the means of most amateurs. However, the advent of sensitive, affordable video cameras has opened a new door for amateurs to participate in these investigations.

Until recent years most occultation observations made by amateurs were performed visually with the aid of simple equipment — a telescope, a stopwatch synchronized with a standard radio time signal (like shortwave station WWV at 5.0, 10.0, and 15.0 MHz in the United States), and in many cases a tape recorder to record vocal call-outs. However, a star's minuscule point of light disappears so quickly that delays in an observer's eye-brain combination and hand-to-stopwatch action or vocal call-out have invariably limited timing accuracies. In the case of grazing occultations, timing accuracies of 0.5 second have long been considered acceptable, while in the case of total occultations, accuracies of 0.2 second or better were required if the data was to be of any real scientific value.

The accurately clocked electronics and rapid frame rates of video cameras provide the perfect solution to the vexing problem of obtaining accurate occultation timings. The pips of the radio time signal can be recorded on a videocassette's audio track. Moreover, videotape recordings can be repeatedly reviewed frame by frame in the playback mode to further reduce uncertainties. The pair of images in Figure 12.2 vividly demonstrates the degree of timing accuracy that can be easily achieved using video. As the Earthlit limb of the Moon occulted a 4th-magnitude star, an interval of only 1/60 second (corresponding to a single video field or semiframe) was all that separated the two successive images. Just try to achieve that level of precision with a stopwatch!

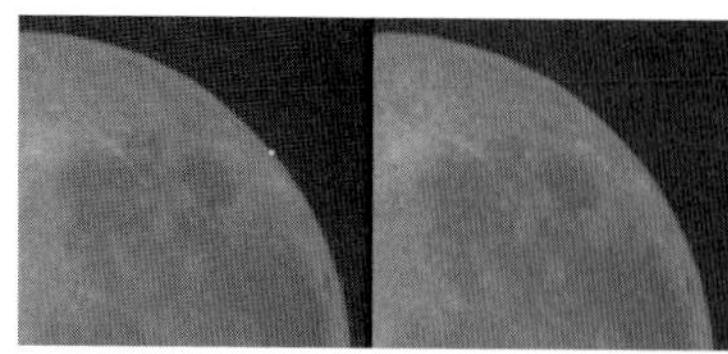

FIGURE 12.2

A 4th-magnitude star disappears behind the Earthlit limb of the Moon. These images are consecutive video fields separated by an interval of 1/60 second.

You don't need a large telescope to get involved with occultation programs (Figure 12.3). Depending on the brightness of the object being occulted, a 60-mm refractor or even a camcorder may be sufficient. You'll also need to know the precise latitude, longitude, and elevation above sea level of your observing site. These coordinates can be obtained from large-scale topographical maps, which are usually available at your local municipal library or can be ordered by mail from the Map Distribution Office, U.S. Geological Survey, P.O. Box 25286, Building 810, Denver Federal Center, Denver, CO 80225. In Canada, contact the Canadian Map Office, 615 Booth St., Ottawa, Ontario K1A 0E9.

FIGURE 12.3

Not an occultation but a very close brush, this dramatic set of images of Saturn skimming past the limb of the Moon was captured by Ron Dantowitz with an Astrovid 2000 camera on the 12-inch Meade Schmidt-Cassegrain at the Boston Museum of Science.

Many local astronomical societies coordinate networks of occultation observers and forward the data they collect to the International Lunar Occultation Center (ILOC) in Tokyo or to the International Occultation Timing Association (IOTA) in St. Charles, Illinois, for further reduction and subsequent reporting.

IOTA has played a major role in encouraging amateurs to employ video for occultation observations. A wealth of useful information on equipment, techniques, reporting forms, and predictions of future events can be found at the IOTA Web site at www.lunar-occultations.com/iota/.

Asteroid Occultations

Occultations of stars by asteroids are also worthy of study using video cameras, though image intensifiers may be required to record many of these events using the modest telescopes available to most amateurs. Multiple timings by coordinated groups of observers have served to determine the dimensions and strange irregular shapes of many asteroids. The presence of unseen satellite companions has even been revealed by unexpected secondary events. If the asteroid is fainter than the star being occulted (and perhaps even too faint to be seen directly or recorded on video through an observer's telescope), the star will simply disappear as the asteroid passes over it. If both the star and the asteroid are sufficiently bright, a decrease in total apparent magnitude of their merged images will occur during the fleeting moments of the occultation. In either case the duration of the event must be precisely recorded.

Amateur collaboration in observing these events is especially valuable because amateurs are mobile. David Dunham, president of IOTA, suggests that even observers not ideally situated on the exceedingly narrow predicted paths of these events make observations nonetheless. His recommendation stems from the fact that many predicted asteroid occultation tracks suffer from inaccuracies of up to hundreds of kilometers due to the imperfectly known orbital parameters of many asteroids. Each year, in the February or March issue of *Sky & Telescope*, Dunham highlights the coming year's asteroid occultations; major events are accorded individual coverage in the magazine's Celestial Calendar department as the time draws near.

Transits of Mercury and Venus

The orbital planes of Mercury and Venus are inclined to the plane of Earth's orbit by angles of several degrees, so at inferior conjunction they usually appear to pass above or below the Sun. But on rare occasions when the lineup is just right, they pass directly across the disk of the Sun and appear as black dots silhouetted against the solar photosphere. Phenomena of this sort are known as *transits* (Figures 12.4, 12.5).

On average, 13 transits of Mercury occur every century. The planet appears as a tiny jet-black spot, darker than the *umbra* (dark central core) of a sunspot. With an apparent diameter of approximately 10 arcseconds, it is too small to make out with the naked eye but makes an easy target for even the smallest telescope.

Transits of Venus are much rarer but far more spectacular than transits of Mercury. They occur in pairs, each pair separated from the next by over a century. The last pair occurred in 1874 and 1882, and the next pair occurs in 2004 and 2012. The disk of Venus in transit subtends an apparent angular diameter of almost one arcminute, well within the grasp of a camcorder equipped with a zoom lens. Even modest telescopes will reveal a host of interesting phenomena caused by sunlight refracted in the planet's dense atmosphere.

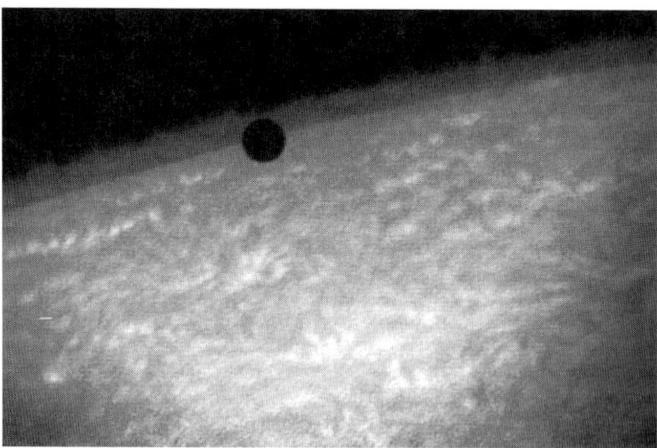

FIGURE 12.4

Ron Dantowitz secured this video image of the November 16, 1999, transit of Mercury through the Big Bear Solar Observatory's 26-inch vacuum reflector, which is equipped with a narrow-bandpass (0.04 nm) hydrogen-alpha filter. Mercury's inky disk is silhouetted against the Sun's photosphere and outer ring of chromosphere.

As dynamic phenomena, transits of the inferior planets are ideal subjects for the astrovideographer. Of course, using a safe over-the-aperture solar filter or Herschel wedge is essential. Narrow-bandwidth (0.15 nm or less) hydrogen-alpha filters are also highly recommended. Alternatively, the image of the Sun can be projected onto a suitable matte white surface and recorded using a camcorder, preferably mounted on a separate camera tripod.

Observers equipped with spectrohelioscopes or hydrogen-alpha filters with a rather broad bandwidth (4 nm or more) should be able to record the interest-

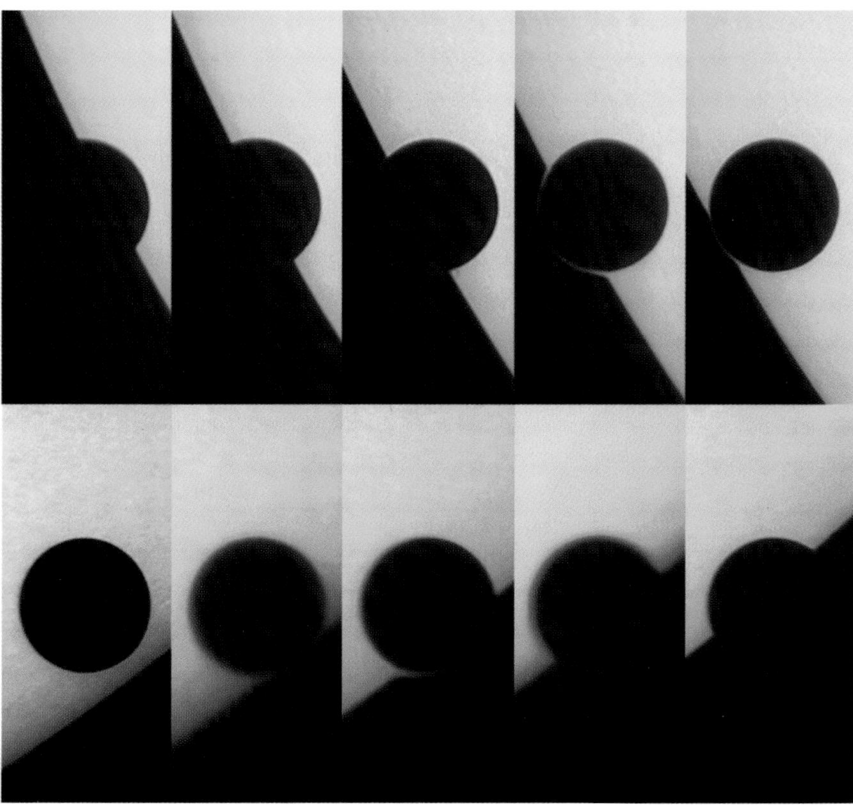

ing spectacle of the disk of a transiting planet projected against the glow of the inner corona for a period of several minutes before or after its passage across the solar photosphere.

FIGURE 12.5
This video sequence of the transit of Venus across the Sun on June 8, 2004, was captured from Tradate, near Milan, Italy, using a Panasonic NV-DS15 digital camcorder and an 8-inch Meade Schmidt-Cassegrain telescope fitted with Baader AstroSolar filter and 40-mm eyepiece. The images show the planet's ingress, or entrance (top row), and egress, or exit (bottom row), across the solar limb. Note the thin ring of light around Venus's silhouette caused by the planet's dense atmosphere refracting light from the Sun.

The Unexpected

It is not unusual for even casual observers to see nocturnal birds, bats, insects, and floating seeds pass in silhouette across the face of the Moon. Occasionally the lunar transit of an aircraft or high-altitude weather balloon is witnessed. Videotapes of these events can be fascinating to review frame by frame in the playback mode.

Predicting lunar transits of artificial satellites can be performed using a variety of readily available satellite-tracking software programs. These events represent challenging targets well worth attempting to record on videotape.

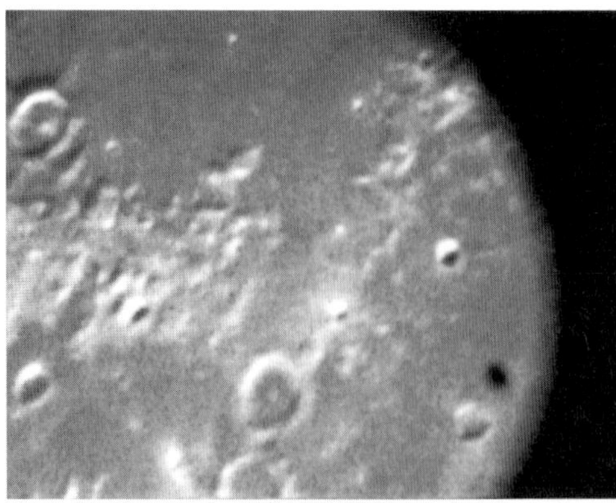

FIGURE 12.6

Unlike the annoyance of an artificial satellite trail appearing on a carefully guided long-exposure photograph of a deep-sky object, this unexpected intruder (lower right) passing across the disk of the Moon remains a minor mystery.

Over the years a number of credible visual observers have witnessed mysterious objects passing across the face of the Moon that are difficult to dismiss as failures to identify birds or bats. In 1896 the famous comet hunter William R. Brooks of Smith Observatory reported the following observation: "While observing the Moon with the 10-inch refractor, a dark round object was seen to move slowly across the Moon in a horizontal direction from E to W. Its apparent diameter was estimated at about one thirtieth of the apparent diameter of the Moon, and the duration of its flight across the Moon was between three and four seconds."

In 1902 the Australian astronomer Sidney Manning was preparing to observe the Moon occult a star when he was startled by a meteor crossing the field of view. A few seconds later, another meteor appeared that followed the same path as its predecessor. Alerted by the passage of the first meteor, Manning was able to follow this object as it transited the face of the Moon. He wrote: "As soon as the meteor was on the Moon it turned black and did not change when crossing the dark plains. It appeared as a perfectly round disk, 30 arcseconds in diameter, and had a granular appearance. The remarkable part of this observation is the meteor's change in brightness."

This intriguing pair of observations is reminiscent of a phenomenon captured on videotape by coauthor Steve Massey on August 8, 1998 (Figure 12.6). A dark circular object subtending an apparent angular diameter of 9 arcseconds rapidly crossed the field. Frame-by-frame playback of the videotape revealed a rate of motion corresponding to an interval of 13 seconds to transit the lunar disk. In all probability the object was a weather balloon adrift at a high altitude, though a meteoroid cannot be ruled out. With a single anomalous observation of this nature, there is an inherent element of doubt as to whether an object is located in Earth's atmosphere or out in space. A handful of astronomers have even attempted to correlate reports like this one with the presence of small, perhaps temporary, natural satellites of Earth, but these notions are very controversial.

The ability to easily derive the dimensions, trajectory, and transit time of this object is further testimony to the power of video as an astronomical tool. If readers manage to capture similar phenomena on videotape — like the overhead passage of orbiting spacecraft (Figures 12.6, 12.7) — they will be able to reexamine their recordings to make a careful assessment. Furthermore, there will be hard evidence of the event rather than mere memories, no matter how vivid.

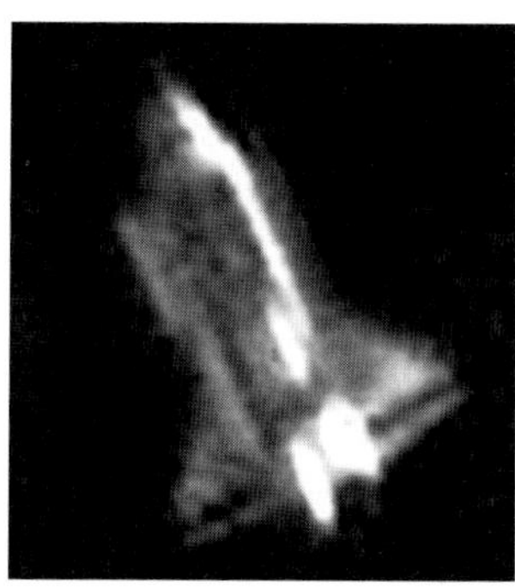

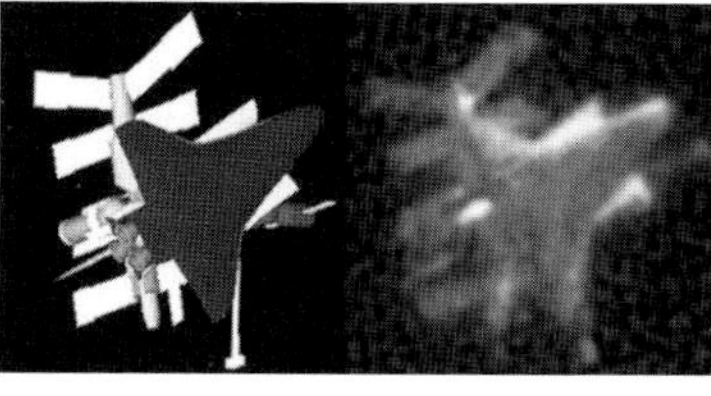

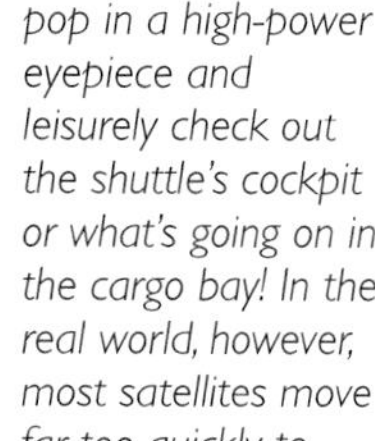

FIGURE 12.7

When imaging orbiting spacecraft, the biggest obstacles are predicting and following their motion. If artificial satellites appeared stationary in the sky, imaging them would be as simple as pointing your telescope and video camera at the Space Shuttle, lining it up in the finder, and looking. In theory you could even pop in a high-power eyepiece and leisurely check out the shuttle's cockpit or what's going on in the cargo bay! In the real world, however, most satellites move far too quickly to keep them centered in an eyepiece's tiny field of view. Try to keep a telescope pointed at the Space Shuttle manually as it flies by at 28,000 km per hour and you're in for a bad case of orbital whiplash! Ron Dantowitz captured these spectacular video images of the orbiting Space Shuttle complete with cockpit windows and cargo bay doors (top image), *and the docking of the Space Shuttle and the Russian space station Mir* (bottom image). *He used a 12-inch Schmidt-Cassegrain riding atop an Archimage mount from Merlin Controls Corporation. With its excellent pointing accuracy and programmable slew rates, the Archimage excels at moving along trajectories that would be utterly impossible to pursue manually. The mount is supplied with superb satellite-tracking software that automatically scans the orbital space above the telescope and displays an ever-changing menu of all the satellites currently visible from your site. Selecting a target from the list causes the mount's DC servo motors to lock on a satellite, and automatically match its apparent angular speed, causing it to stand still through an eyepiece or on a CCD chip.*

Eclipses

13

Solar Eclipses

No astronomical phenomenon is more spectacular than a total eclipse of the Sun. While conventional photography can capture much of the eerie beauty of these events, a dynamic medium like video is a convenient way of recording the unfolding drama. The remarkable compactness of camcorders combined with their zoom optics make them the ideal tool for eclipse chasers (Figure 13.1).

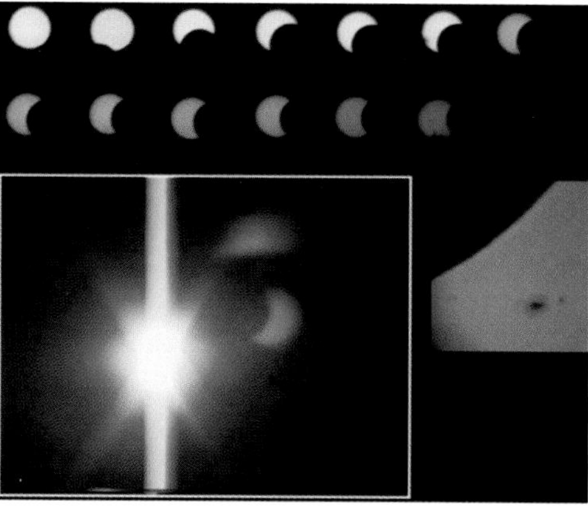

FIGURE 13.1
The partial solar eclipse of February 16, 1999, from Sydney, Australia. The sequence of images at the top was captured using a tripod-mounted camcorder equipped with a solar filter and 8× zoom lens. Bottom left: *At mideclipse the solar filter was removed and the shutter speed set to 1/2000 second.* Bottom right: *An image obtained by holding the camcorder to the eyepiece of a telescope equipped with a solar filter.*

Image Scale

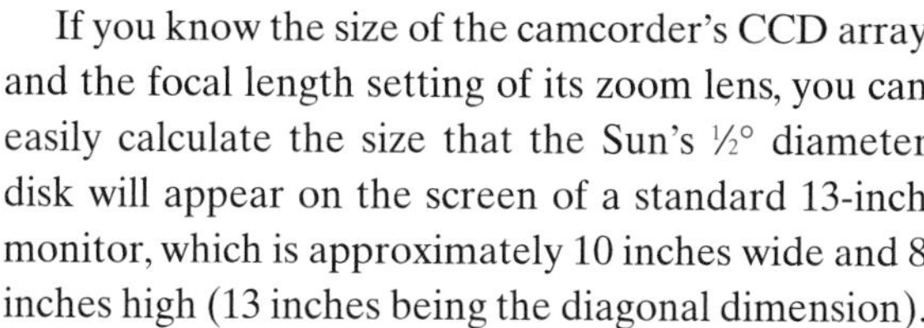

Most camcorders are equipped with zoom lenses providing magnifications up to about 22×. You'll want to employ only the optical zoom powers of your camera, not the digital zoom feature found in many cameras.

If you know the size of the camcorder's CCD array and the focal length setting of its zoom lens, you can easily calculate the size that the Sun's ½° diameter disk will appear on the screen of a standard 13-inch monitor, which is approximately 10 inches wide and 8 inches high (13 inches being the diagonal dimension).

Usually the focal length range of a camcorder's zoom lens is marked in millimeters. The CCD arrays in most camcorders range from ¼-inch to ⅔-inch format. If in doubt, check the manual. The Sun's image displayed on a 13-inch monitor will be equivalent to the focal length setting of the camcorder's zoom lens multiplied by 0.80 for a ¼-inch format array, 0.60 for a ⅓-inch format array, 0.40 for a ½-inch format array, or 0.34 for a ⅔-inch format array. For example, a camcorder with a ½-inch format array and a 65-mm focal length lens will display an image of the Sun about 26 mm (about 1 inch) across on a 13-inch screen. If you're accustomed to thinking in terms of image scale relative to a frame of 35-mm film, you can convert the focal lengths of camcorder lenses to their approximate 35-mm equivalents by multiplying them by 10.4 for a ¼-inch chip, 9.0 for a ⅓-inch chip, 5.2 for a ½-inch chip, or 4.5 for ⅔-inch chip.

A 1-inch diameter solar image is fairly small, but will still yield an acceptable record of the diamond ring and corona. With an effective focal length of 200 to 400 mm and a ½-inch format chip, the Sun's image will appear about 3 to 6 inches across on a 13-inch monitor. You don't want to fill the screen with the Sun's disk because the solar corona extends more than one solar diameter on either side of the Sun.

In the months preceding a solar eclipse, you can use the Moon as a convenient subject for practice runs with your camcorder because the apparent angular diameter of the Moon and the Sun are virtually identical. Moreover, the full Moon is comparable in brightness to the inner corona.

Another factor to consider is the distance the Sun will travel across the sky during totality. Since the Sun and Moon appear to move across almost 15° of sky every hour, they will travel a full disk diameter during each minute of totality. If possible, use a small motor-driven equatorial mount to support your camcorder rather than a camera tripod with a pan head, particularly if you plan to zoom in for detailed views of prominences and Baily's Beads.

Finally, some camcorders come with hand-held remote controls. These are a real boon to eclipse videographers because they allow you to operate the zoom lens and change exposure settings without jiggling the image. Figures 13.1 and 13.2 were shot using camcorders. Figures 13.3 and 13.4 were taken with Astrovid cameras.

Partial Phases

As with conventional photography, a solar filter must be placed over the lens of the camera to record the partial phases of a solar eclipse. A #14 welding filter is inexpensive, readily available, and works well, but gives a yellow-green solar image. The solar filters that consist of an evaporated metal film on a glass substrate provide pleasing yellow or pale orange solar images. The popular SolarSkreen Mylar filters provide a cool blue image, but if they are combined with a Wratten #85 salmon filter the solar image will be virtually colorless.

A question asked by many prospective eclipse videographers is: "When should I remove the filter?" Momentary exposure to full sunlight intensity could irreparably damage or even destroy the sensor of an old vidicon tube camera in the twinkling of an eye, but today's solid-state CCD detectors are more or less impervious to overexposure. You can remove the filter at the onset of the first diamond ring and not replace it until five seconds or so after the second diamond ring. The diamond ring effect occurs for a fraction of a second before the Moon completely obscures the Sun at the beginning of totality (second contact) or at the end of totality as the moving Moon reveals the Sun.

Totality

No filter is needed during totality itself, but because of the high sensitivity of some cameras it may help to use a small amount of filtering — say a neutral-density 0.9 filter — especially if you are interested in recording structure in the prominences and extreme inner corona. The automatic gain control of most cameras will be all the way up with or without the filter, since the overall field is relatively dark. At maximum gain the bright inner corona will probably be overexposed.

FIGURE 13.2
Imelda B. Joson used a tripod-mounted Canon XL1 Digital video camera with a 16x zoom lens to record this image of the inner corona and several bright prominences during the August 11, 1999, solar eclipse from Harput, Turkey.

If you have a manual override on your camcorder's automatic gain control, you can adjust it to suit your tastes while watching the image through the viewfinder or on a separate monitor. Another trick for recording the features of the inner and outer corona is to vary the electronic shutter speed provided that your camcorder has a manual shutter control.

Be sure to turn off the camera's autofocus feature and manually set the focus to infinity. If the autofocus function is not disabled, the camera may continually hunt for best focal setting, repeatedly moving in and out of focus. This will prove very annoying during the precious moments of totality!

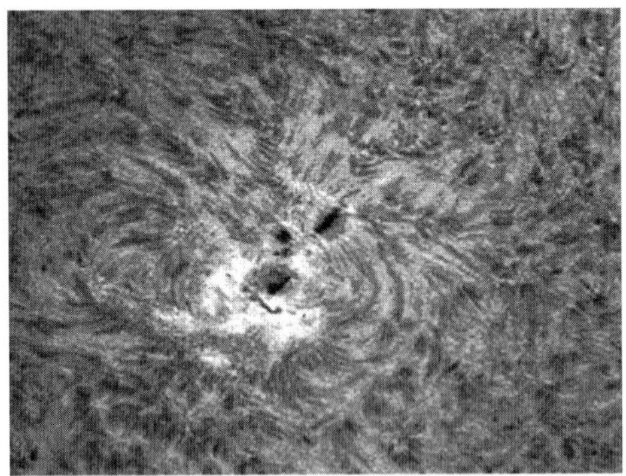

FIGURE 13.3
Even without the benefit of an eclipse, with proper filtration the Sun is an excellent subject for video cameras. Gordon Garcia recorded this sunspot on October 11, 1998, using a Daystar hydrogen-alpha filter and an Astrovid 2000 video camera on a 5-inch refractor. The image was captured in real time to a computer using a Snappy frame grabber.

Many cameras can also display the time in the field of view of the recording, but most videographers are after maximum aesthetic appeal and prefer unsullied images. The camcorder's audio channel is perfect for adding notes during the event or just recording the hoots, howls, whistles, laughter, and excited utterances of people around you.

You might consider the luxury of a second camcorder with a wide-angle lens atop a fixed tripod to record the sweep of the Moon's shadow as it races toward the site, or to record any shadow bands that may be visible.

Lunar Eclipses

The Moon subtends an apparent angular diameter of ½°, so the calculations of focal lengths and the corresponding image scales provided in the section on solar eclipses apply here as well. The light sensitivity of modern camcorders is adequate for recording the partial phases of a lunar eclipse, but it may be necessary to reduce the image scale during totality. A fixed camera tripod or other sturdy support is essential. Once again, it is preferable to mount a camcorder atop an equatorial mount equipped with a motor drive or piggyback it on a telescope.

Closed-circuit surveillance cameras are ideal for recording lunar eclipses, particularly when they are combined with the telephoto lenses made for 35-mm format film cameras using a C- to T-thread adapter. The optimum focal length of the telephoto lens will depend on the size of the camera's CCD array sensor. You'll probably want the image of the lunar disk to not quite fill the frame, leaving a bit of sky all around it. To nicely frame your subject, the various chip formats and the corresponding optimal focal lengths are as follows: for a ¼-inch chip, 200 to 250 mm; for a ⅓-inch chip, 300 to 350 mm; for a ½-inch chip, 400 to 500 mm; for a ⅔-inch chip 600 to 700 mm. There's plenty of light available for recording the partial phases, but you'll want to work at f/2.8 to f/4 to record the ruddy hues during totality, especially if the eclipse is a dark one due to the presence of volcanic dust and aerosols suspended high in Earth's atmosphere.

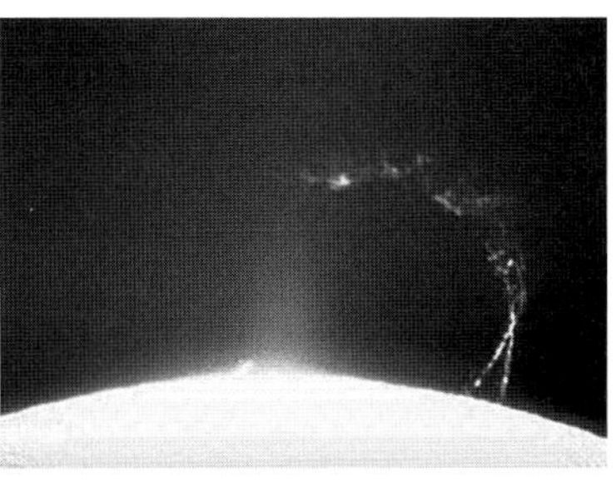

FIGURE 13.4
This image of a solar prominence is a composite of three frames grabbed with a Snappy from an S-VHS videotape recording with the Astrovid 2000 camera.

Byron Soulsby of the Calwell Lunar Observatory in Australia produces video images from a Sony Handycam TR511E camcorder to broadcast lunar eclipses live via the Internet. He simply piggybacks his camcorder on an equatorially mounted 6-inch Newtonian reflector that is equipped with a motor drive. Since the camcorder can obtain clear pictures of the Moon at 36× using its zoom lens in conjunction with a 2× teleconverter, its electronic zoom feature is not employed.

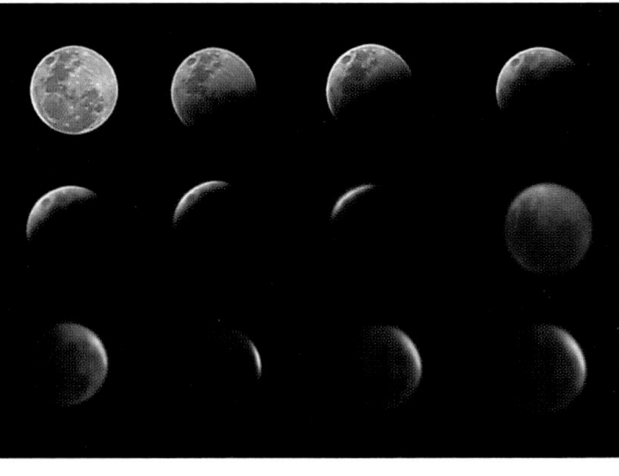

FIGURE 13.5
A montage of the total lunar eclipse on July 17, 2000, created from images obtained with an ordinary Sony camcorder equipped with a zoom lens.

The camcorder's video output signal is fed to a Macintosh (IIci) computer equipped with a 24-bit color frame grabber. Images are processed with a networked PowerMac and saved as GIF or JPEG format files. In preparation for an eclipse Soulsby creates a new Web page for the event with a sample image window. At intervals of two or three minutes during the eclipse, new images are sent to the Web site using an FTP (File Transfer Protocol) program.

Following the eclipse, a sequence of the entire event is created using GIF animation software. This file is then posted to the same Web page window where images had been displayed during the eclipse. During the total lunar eclipse of September 16, 1997, Soulsby's Web site received almost 55,000 hits!

The Deep Sky 14

A Video Guidescope

If you've ever manually guided a long-exposure photograph of a deep-sky object, you're probably all too aware of just how sore your shoulders and neck can get. Keeping an eye glued to an illuminated-reticle eyepiece mounted in an off-axis guider or a piggybacked guidescope requires intense concentration, often while maintaining an awkward pose for prolonged periods of time.

A compact video camera and a monitor can help you avoid those aches and pains and provide the ultimate in relief from eye strain. In fact, in today's marketplace a suitable low-light monochrome video camera costs no more than an illuminated reticle eyepiece!

Set up your telescope as you would normally for conventional photography. Once you've visually acquired an appropriate guidestar, remove the eyepiece and insert the video camera in the focuser of your guide telescope or off-axis guider.

A Barlow lens or eyepiece projection may be required to provide a sufficiently large image scale for precise guiding. An old rule of thumb among astrophotographers is that the guiding magnification should be no less than five times the focal length in inches (or twice the focal length in centimeters) of the telephoto lens or telescope taking the photograph. As an example, let's take the case of a 10-inch f/4.5 Newtonian with a focal length of 45 inches. A minimum magnification of $5 \times 45 = 225\times$ is required to prevent trailed stars on a photograph taken at the prime focus of this telescope.

When the image displayed on the video monitor is compared to the view through a conventional guiding eyepiece with a 45° field of view (typical of the common Kellners and orthoscopics), the size of the video camera's chip will correspond to the focal length of the eyepiece. A ½-inch format chip is equivalent to a 7-mm eyepiece, a ⅓-inch format chip to a 5-mm eye-

FIGURE 14.1

A compact, lightweight video camera has been mounted at the focus of a 60-mm refractor guidescope, while a 35-mm SLR film camera for deep-sky photography is located at the prime focus of a 10-inch f/4.5 Newtonian reflector. The video monitor's screen has been marked with a cross hair to serve as a guiding "reticle."

piece, and a ¼-inch format chip to a 3.5-mm eyepiece. In the case of a typical guidescope like a refractor with an aperture of 80 mm and a focal length of 1,200 mm, the equivalent magnification obtained with a ⅓-inch format chip at its prime focus is 1,200/5 = 240×, adequate for guiding a photograph through the 10-inch f/4.5 reflector in this example. A shorter focal length guidescope would require the use of a Barlow lens or eyepiece projection to achieve an adequate image scale.

Used in conjunction with a video camera rated at 0.05 lux, an 80-mm aperture guidescope gathers sufficient light for guidestars of about 6th magnitude. A 60-mm aperture guidescope will require a guidestar of 5th magnitude or brighter.

Your monitor will now serve as your guiding "eyepiece" (Figure 14.1). Make a removable cross hair on the screen using thin strips of tape or a felt-tip marker.

Focus the video camera–guidescope combination. Then, with a few touches of the buttons of your dual-axis drive corrector, determine the directions of motion in right ascension and declination on the monitor. Rotate the video camera so that motions in right ascension and declination correspond to the axes of your cross hair. Now you can correct the motions of the guidestar while comfortably seated a few feet in front of the monitor instead of hunching over and squinting to peer through a short focal length eyepiece.

Extreme Video

Without optical aid or through a modest set of

binoculars, our eyes can readily see the bright cloud of hydrogen that surrounds young stars in M42, the great Orion Nebula. But combining even a small telescope with photographic film or CCD camera and an exposure lasting several minutes will reveal a wealth of structural details and features in the nebula that are all but invisible to the eye.

FIGURE 14.2
The Trapezium and two other stars in the Orion Nebula as they appear on a standard video monitor.

Standard video cameras, limited by their relatively short exposures, fare considerably worse than the human eye when it comes to detecting faint sources of light. Even under light-polluted suburban skies, a visual observer armed with a 10-inch reflector will see striking views of M42's intricate swirls of nebulosity. But put a 0.05-lux video camera at the focus of this telescope and barely a hint of nebulosity will appear on the TV monitor, only a sprinkling of the nebula's brightest stars (Figure 14.2).

Until recently, the only way to view and record faint nebulae, comets, or galaxies was to use an *image intensifier*. This device can provide memorable telescopic views from light-polluted urban or suburban locations, where museums, planetariums, and public observatories are typically located.

The heart of a state-of-the-art image intensifier is a cylindrical vacuum tube, typically about 18 or 25 mm in diameter. Special semiconductor layers at the tube's front end act as a photocathode, converting the incoming photons into electrons. A modest gate-voltage field is applied to the photocathode to accelerate the emitted electrons onto a microchannel plate. The plate is a disk of special glass perforated by millions of closely packed holes, or microchannels. The glass is slightly conductive so when voltage (typically 800 to 1,000 volts) is applied, it forms a constant field along the length of the microchannels. This causes the electrons to accelerate through the holes and produce a cascade of secondary electrons as they ricochet off the microchannel walls. After emerging from the plate, an even higher voltage further accelerates these

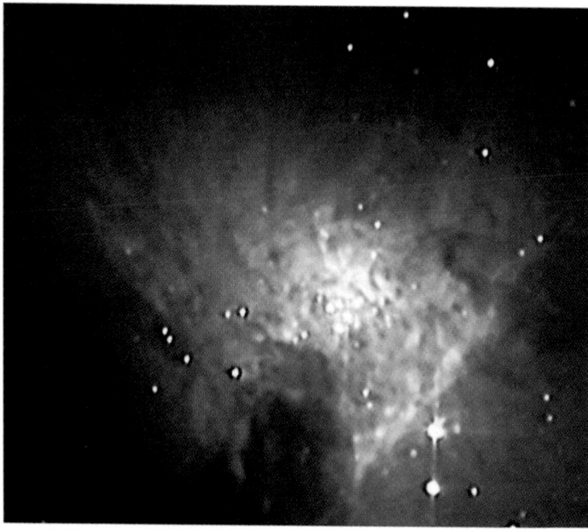

FIGURE 14.3

These tricolor images of the Orion Nebula, M42 (top), and the galaxy M83 in Hydra demonstrate the amazing image-integration capabilities of today's highly sensitive, "uncooled" video cameras. Both pictures were taken at the focus of an 8-inch Newtonian reflector.

secondary electrons onto a phosphor screen, producing a highly amplified replica of the incoming light's original image.

When struck by electrons, the phosphor screen emits green monochromatic light that closely matches the peak sensitivity of the human eye. The most advanced image-intensifier design, known as "generation-three" devices, can amplify light 30,000 to 100,000 times, making them ideal for use in night-vision scopes. When coupled to a sensitive video camera (via a relay lens), an image intensifier can deliver very good images of deep-sky objects.

In recent years, however, image intensifiers have fallen somewhat by the wayside since the introduction of a new breed of extremely sensitive video cameras with onboard frame-integration capabilities. These units boast sensitivity ratings from 0.0001 to 0.00005 lux at f/1.2. The Mintron 12V1-EX, for example, uses Sony's EXview HAD CCD chip and is one of the most popular models being employed today for deep-sky imaging and meteor surveillance. While these devices operate like ordinary security cameras when imaging bright targets such as the Moon and planets, they offer, among other features, frame-integration modes with increments from 2× to 128×. Incoming images are stored and accumulated internally, building up a picture that is refreshed at a slower rate with each increasing incremental step. For instance, at the 128× setting, the picture is refreshed roughly every 2.6 seconds. When combined with fast, wide-angle lenses, stunning views of the sky can be recorded in amazing detail. This major evolution in video-camera technology is also being used for detecting comets and even supernovae down to 16th magnitude with 8- to 10-inch telescopes!

A minor drawback in some of these cameras when using long integration modes is the presence of "hot" pixels in the image. To remove these artifacts from the final picture, during image processing you need to digitally subtract a "dark frame" (a blank image taken with the telescope covered and using the same exposure settings as those used to record the subject). Several processed images can also be stacked together to produce a smoother, more aesthetically pleasing composite.

FIGURE 14.4
The I^3 Piece image intensifier (far left) from Collins Electro Optics with a relay lens and closed-circuit GBC-500 video camera, mounted at the focus of a 16-inch Schmidt-Cassegrain.

FIGURE 14.5
A Sony camcorder coupled to an image intensifier, ready to be attached to the focuser of a telescope.

Real-Time Frame-Adding Software

While a single frame of a bright subject like M42 taken with a standard 0.05-lux security camera and a 10-inch telescope will show virtually zero nebulosity on the monitor, in reality the camera has indeed captured some of the nebula's faint light. We just can't see any trace of it because its low-contrast details are lost in the image's random background noise. Using your computer's video-capture port, processing software such as COAA's *AstroVideo* can help amplify the faintest signals registered by the camera's CCD sensor. It accomplishes this by coadding the incoming video frames, thereby producing a final picture with greatly improved signal-to-noise ratio. The picture's contrast can then be stretched to reveal its otherwise hidden details.

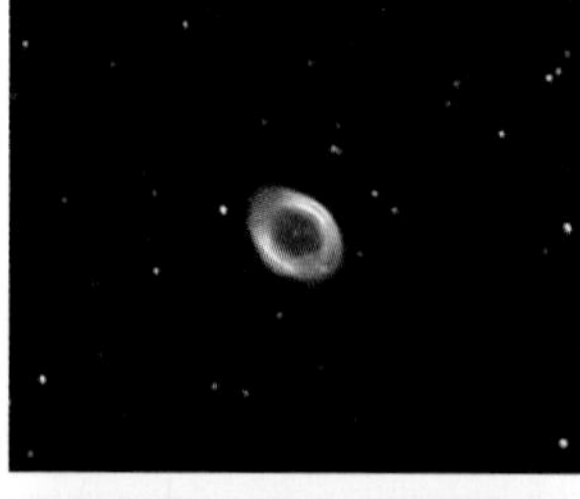

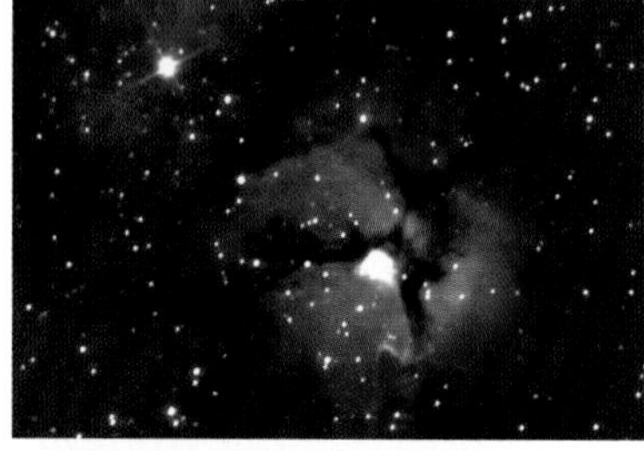

FIGURE 14.6

A low-cost alternative to military-grade image intensifiers for imaging deep-sky objects are the new generation of frame-integrating video cameras. These tricolor portraits (from top to bottom) of NGC 253 (Sculptor Galaxy), M57 (Ring Nebula), M20 (Trifid Nebula), and NGC 5128 (Centaurus A) illustrate the amazing capabilities of the Mintron MTV-12V1C-EX video camera when coupled to a 10-inch f/5 Newtonian reflector. The images were taken through red, green, and blue (RGB) filters under light-polluted skies in the suburbs of Sydney.

Meteors

15

Meteor Surveillance with Video

Professional astronomers in Japan, Germany, and The Netherlands pioneered the use of video to study meteors in the 1960s and 70s. The results were very encouraging, but the cost of their video systems was prohibitively high.

Members of the International Meteor Organization (IMO), an amateur network founded in 1988, faced a similar problem — the high cost of constructing suitable video systems forced many to just pool their resources in order to construct a single camera. This situation persisted until the mid-1990s, when steadily dropping prices and improving performance of second- and third-generation image intensifiers substantially reduced production costs, greatly broadening video's appeal to an increasing number of amateur meteor specialists.

IMO's later camera designs became more compact, featuring modular components mounted on a spar. A fast (f/1.2 to f/2.8) objective lens of 20 to 50 mm focal length coupled to a second- or third-generation image intensifier amplifies incoming light by a factor of 10,000 to 100,000. A high-quality, flat-field relay lens focuses the image formed on the intensifier's phosphor screen onto the CCD array of a sensitive, black-and-white security video camera. At a cost of several thousand dollars, homemade image-intensified video systems can record meteors as faint as 9th magnitude within a 20° field. They offer a number of important advantages over traditional observing methods, among them:

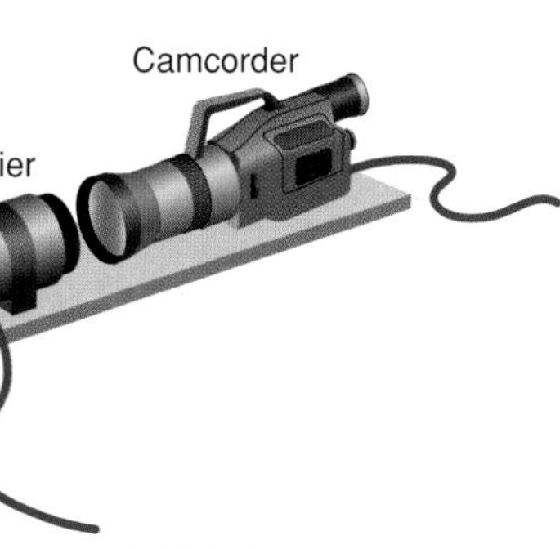

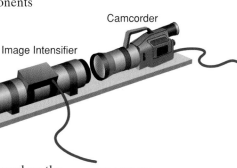

FIGURE 15.1
A schematic of the modular, spar-mounted meteor camera developed by the International Meteor Organization. The image intensifier is equipped with a fast, wide-angle lens, while the camcorder operates in "macro" mode to record the image intensifier's output on the phosphor screen.

1. Meteors too faint to be seen visually or recorded by conventional photographic emulsions can be recorded, supplementing observations with radio and

radar. Moreover, the field of view of these systems can be comparable to that covered by a visual observer.

2. The light curve (brightness profile) of meteors can be recorded, providing important information about the properties of the parent meteoroids and how they interact with Earth's upper atmosphere.

3. Video provides an objective record of changes in the intensity of meteor showers over time.

4. Last, the angular speed and velocity of a meteor can be determined directly.

Scanning hours of videotape recordings to search for meteors was a very tedious and inefficient process. This all changed in 1993, when German meteor enthusiast Sirko Molau introduced his *MetRec* software, which automated the process. The most recent version of *MetRec* requires a computer with a minimum of 166 MHz or faster processing speed, a video-capture card, and at least 16 MB of memory. At the time of writing, it runs only in DOS, such as the DOS mode in *Windows 95* and *98* operating systems. After the video's input signal is digitized, the program subtracts pairs of successive frames to make stars and other stationary objects disappear. Next, the resolution of each frame is reduced by a factor of four by averaging 16-pixel luminance values and converting the frames from 512-pixel-square to 128-pixel-square. This algorithm reduces the image's background noise by 75 percent. Then, a "mask" frame representing the chip's average dark-current noise is subtracted from each frame. Finally, *MetRec* looks for linear features above a predetermined brightness threshold.

Serendipitous video recordings of fireballs, including a few that have resulted in meteorite falls, are often the only data available that permitted astronomers to determine the orbits of these objects. Commonly utilized by organized fireball patrols, video has also provided valuable insights on the way these bodies break up as they plunge through Earth's atmosphere.

Amateur fireball patrols carried out in Australia in the late 1990s used six all-sky photographic cameras stationed at six different locations. But timing the beginning and end of a fireball event could not be accurately determined from a long-exposure photograph, so some film cameras were replaced with sensitive black-and-white security video cameras for continuous monitoring. The videotapes were stored until the patrol photographs could be developed and examined. If a fireball appeared on a photograph, the videotapes were then reviewed to determine the exact times of the event. If not, the tape was reused.

Acquiring a fish-eye lens with an extremely short focal length that would cover the entire sky with a typical black-and-white video camera (equipped with a tiny ⅓-inch-format chip) can be a challenge. Some amateurs resort to pointing such a video camera (fitted with a "normal," or standard, lens) downward at a convex mirror and recording the reflected image of the sky. They had to carefully match the focal length of the video camera's lens and the height of the camera above the convex mirror so that the system's field of view spans 180° of sky (horizon to horizon). An illuminated digital watch placed in one corner of the frame served as time reference. (Alternatively, a reference time signal broadcast by a radio station, say, WWV in the United States, could be fed into the audio input of a VCR.)

FIGURE 15.2
When fitted with a 2.6-mm lens, the Mintron MTV-12V 1-EX video camera yields a field of view spanning 85° by 65°. Under dark skies and using the 2× frame-accumulation mode, this camera can record stars down to around 3rd magnitude, perfect for catching bright meteors. This fireball achieved a peak brightness of about magnitude – 4.

Using video for positional measurements was not suitable. A video image typically has only 540 pixels to cover the sky from horizon to horizon, which represented a resolution of only a third of a degree (20 arcminutes) per pixel — about an order of magnitude lower than the resolution of photographic all-sky cameras. In addition, a video camera with a sensitivity of 0.05 lux and equipped with an f/1.4 lens recorded only the brightest handful of stars reflected by the convex mirror, providing very few reference points with which to determine a fireball's coordinates. Finally, distortions introduced by the camera's optics also had to be taken into account, confounding the setup's already-limited positional accuracy. For these reasons, video served only as a source of time data for meteors.

The brightness of a fireball (usually defined as a meteor approximately as bright as Venus, or about magnitude –4) is at least two orders of magnitude

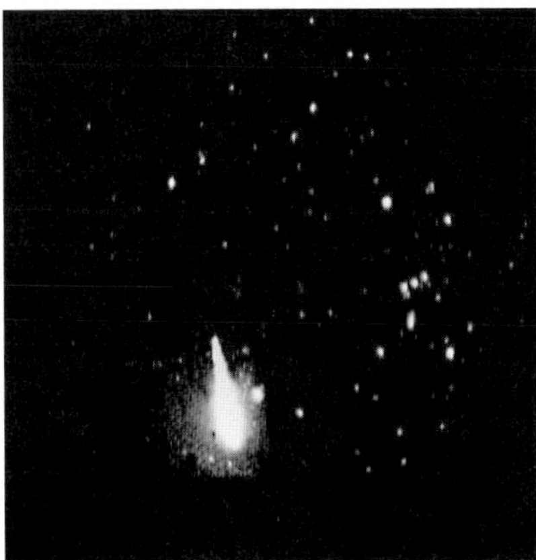

FIGURE 15.3
This Delta Cancrid fireball was recorded on the night of January 5–6, 1995, over Hannover, Germany. The video frames showing the meteor were digitized, processed, and converted into an MPEG animation.

brighter than the threshold of a typical video system, so fireballs can be picked out easily on the monitor when the tape is played back at 10 times the normal viewing speed. In addition to these natural intruders, it's also possible to record man-made space debris, such as spent rocket boosters or dead communications satellites, as they re-enter the atmosphere and burn up.

Since the primary purpose of collecting video data is to accurately determine the times of a fireball event, the quality of the recorded images is not of paramount concern. Consequently, there is no need to employ the VCR's "standard-play" mode when recording. Using the "long-play" or "extended-play" mode with a 2-hour videocassette tape will provide up to 6 hours of recording time. (You would still need to change the tape when recording during long winter nights.) An all-sky video camera could actually be operated 24 hours a day if it is equipped with a motor-driven occulting disk to block the Sun. Although daylight fireballs are reported much less frequently than their nocturnal counterparts, the video record in such an instance could be used for positional measurements.

A number of automated fireball detectors have been developed over the years, many of which have employed photomultiplier tubes as sensors. More recently, highly economical and equally sensitive surveillance systems are being designed around the new generation of integrating video cameras. These cameras, used in conjunction with Sirko Molau's wonderful *MetRec* software, have recorded unprecedented numbers of bright and very faint meteors.

In addition to video data, fireballs are the only celestial phenomena capable of producing audio data. When a sizable meteoroid enters Earth's atmosphere at hypersonic speeds, it generates a powerful shock wave. In the extremely rarefied air at high altitudes, this shock wave is refracted away from Earth. But at altitudes of less than 50 kilometers it can produce a very loud, cracking sound, which can be heard from the area directly under the fireball's path and for a short distance around it. Known as "primary compression wave," it is essentially the same as a sonic boom produced by an aircraft flying at more than the speed of sound. This shock wave can be extremely loud, and has been known to break windows as well as frighten people and livestock.

A prolonged low-frequency rumbling that resembles the sound of a rocket being launched may be heard over a much wider area. These rumbling noises have been known to persist for more than a minute and may represent echoes of the primary shock wave. An omnidirectional microphone should be capable of recording both types of sounds on the tape's audio track.

Meteorite Impacts on the Moon

Over the years, a few credible visual observers have reported transient flashes of light on the Moon that were widely interpreted as meteorite impacts. Here are a few typical examples:

October 19, 1945: Using a 9-inch Newtonian reflector at a magnification of 220×, British selenographer F. H. Thornton saw a brilliant orange-yellow point of light on the floor of Plato near the ramparts of the crater's eastern wall. With impressions of the Second World War fresh in his mind, Thornton compared the phenomenon to "the flash of an antiaircraft shell exploding in the air at a distance of about ten miles." He noted that the date was in the middle of the Orionid meteor stream.

April 15, 1948: A. W. Vince was examining the Earthlit portion of the crescent Moon with a 6.3-inch refractor when he was startled by a momentary flash, similar in brightness to a 3rd-magnitude star, located near the darkened limb some 30° north of the crater Grimaldi, which he was able to distinguish as a dark patch.

August 8, 1948: While examining the Earthlit part of the Moon at a magnification of 50×, A. J. Woodward saw a flash "like a bright sparkle of frost on the ground" that at first appeared bluish white, then turned yellow and faded from view after about three

FIGURE 15.4

This image of the impact of a Leonid meteor on the Earthlit portion of the Moon was obtained by David Palmer with a 0.03-lux monochrome video camera on a 5-inch Schmidt-Cassegrain operating at f/6.3. It is a composite of 10 video fields.

seconds. To Woodward it had the appearance of an object striking the Moon's surface.

April 24, 1955: F. C. Wykes was observing the Moon when he saw a white flash of short duration on the Earthlit portion of the disk in the northern part of Mare Serenitatis, not far from the crater Posidonius.

Given the very brief duration of these events, it is hardly surprising that only a single supporting piece of photographic evidence was ever obtained, a snapshot through an 8-inch reflector taken by Lyle Stuart in 1953 that showed a bright point of light near the terminator of the first-quarter Moon. It was only in 1999 that video observations would finally vindicate generations of visual observers.

David Dunham, president of IOTA, organized a network of amateur astronomers to monitor the Earthlit portion of the Moon during the 1999 Leonid meteor shower. The Leonid meteors, so-named because they appear to radiate from the constellation Leo, are the debris shed by Comet Tempel-Tuttle. These particles move through space in streams that are pulled in different directions by the gravitational attraction of the larger planets. Calculations by David Asher of Armagh Observatory in Northern Ireland and Rob McNaught of Siding Spring Observatory suggested that Earth would pass through a particularly dense part of the Leonid stream on the night of November 17–18, 1999. The Moon was also well placed at that time.

FIGURE 15.5

A second image of an impacting Leonid meteor by David Palmer.

Equipped with sensitive (<0.05 lux) black-and-white video cameras and a variety of small telescopes, Dunham's network of observers recorded WWV time signals on the audio track of their videotapes so that any suspicious flashes could be identified and independently confirmed. On videotapes recorded by more than one observer, six flashes appeared at the same instant and at the same location on the Moon, compelling evidence that they represented genuine lunar impacts. Ranging in brightness from 3rd to 7th magnitude, they were all located just north of the lunar equator on or near the lava plain known as the Oceanus Procellarum. Each impact appeared bright on only a single video frame, and was recorded at a considerably fainter level on the subsequent frame that followed 1⁄30 second later (Figures 15.4, 15.5).

These successful results with surprisingly modest equipment will surely encourage video observers to record the Moon during future meteor showers. The desirability of recording a substantial fraction of the Earthlit portion of the Moon dictates that the focal length of the telescope must be rather modest, so refractors and Newtonian reflectors of 4 to 6 inches (10 to 15 cm) aperture with focal ratios of f/4 to f/7 should be ideal instruments. Dunham provides the following useful advice for the owners of popular catadioptric telescopes: "A key to my success in this endeavor was using a focal reducer (telecompressor) that decreased the focal ratio of my 5-inch Celestron Schmidt-Cassegrain from f/10 to f/6.3. That not only increased the field of view by more than a factor of 3 in area, but also increased sensitivity by concentrating the seeing disk of point sources of light onto fewer pixels, as well as allowing the faintly Earthlit portion of the gibbous Moon to be recorded."

Time-Lapse Animation 16

A time-lapse movie is made by assembling a sequence of still images of an event that occurs slowly, like the sprouting of a seed, the opening of a blossom, or the construction of a skyscraper. The individual stills, taken at intervals of minutes, hours, or even days, are spliced together and rapidly displayed so that a viewer sees an event appear to miraculously unfold in a matter of seconds.

Time-lapse animations assembled from video recordings make dramatic displays of many solar system phenomena. Suitable subjects include the advance and retreat of shadows across the lunar landscape (Figure 16.1); the month-to-month changes in the librations of the Moon; the waxing and waning of the phases of Venus (Figure 16.2); the evolution of solar prominences in hydrogen-alpha light; the axial rotation of the Sun, Mars (Figure 16.3), and Jupiter; the eclipses, occultations, and transits of Jupiter's Galilean satellites; and occultations of stars and planets by the Moon.

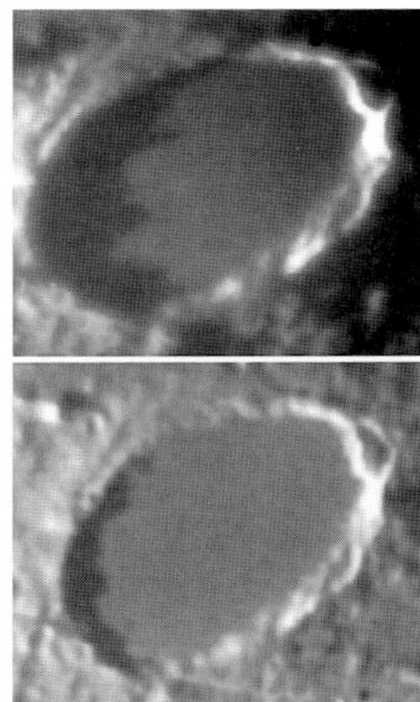

FIGURE 16.1

Sunrise over the lunar crater Plato is an excellent subject for a time-lapse animation. Note the pointed spire of shadow extending across the crater's smooth, dusky floor in the upper image.

FIGURE 16.2

The changing phase of Venus from August to October 2002 was captured in this sequence with an Astrovid 2000 video camera and a 10-inch Newtonian telescope. Most of the images were recorded during daytime.

Planning

Always consider your target audience. How long will the subject hold their interest? Will they want to see subtle incremental changes or rapid, dramatic ones? These factors will help to determine the capture rate and the length of the final movie. Instead of the customary video capture rate of 25 to 30 frames per second, you may want to produce a movie that consists of frames taken at one-minute intervals. When such a

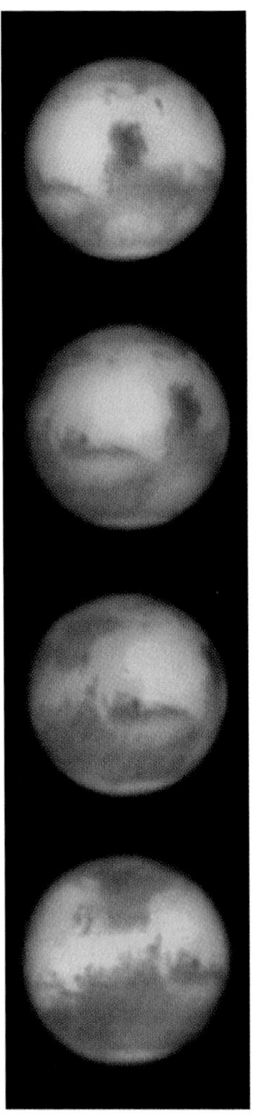

FIGURE 16.3

These frames are from a time-lapse animation of the rotation of Mars in 2001 recorded with the 24-inch telescope in Siding Spring, Australia. The dark wedge-shaped feature situated on the planet's central meridian in the top image is Syrtis Major. A thin white cloud can be seen extending to the east across the Tyrrhenum and Cimmerium regions as well as along the Martian limb.

sequence is played back at a rate of 10 frames per second, the rate at which an event appears to occur is accelerated by a factor of 600.

You'll need to choose an appropriate event to document. If you have decided to capture the retreating shadows as the Sun rises over a lunar landscape near the terminator, you'll probably want to select an imposing crater with terraced walls or a prominent central peak. Isolated mountains like Mount Pico and Mount Piton are also excellent subjects for recording the effects of changes in lighting.

If you have decided to record the transit of a Galilean satellite or its shadow across the face of Jupiter, determine the starting and ending times of the event by consulting an ephemeris. A variety of computer programs can be particularly helpful here. Some celestial events are rather brief (the transits of Io and Europa) while others are comparatively long (the axial rotation of the Sun or the changing phases of Venus), so careful planning is required to capture a sufficient number of frames so that the transitions between frames appear smooth and continuous when the movie is viewed.

Centering the Images

It is vitally important that you keep your subject accurately centered during each recording or capture sequence. Use a felt-tip marker or strips of narrow tape to make a simulated cross hair reticle (an X) on the screen of the monitor. Record or capture only when the subject is precisely centered with respect to this point of reference.

When lunar occultations are recorded at a moderately large image scale, the Moon's limb will appear as a gently curving arc on the monitor. Use a felt tip marker to trace this curve on the screen. Once the proper adjustments to the telescope's tracking rate have been made, only an occasional tweak of the slow-motion controls will be required to keep the lunar limb coincident with this reference line. If the occultation occurs at the dark limb of the Moon that is illuminated only by earthshine, careful adjustments to the monitor's brightness and contrast controls will be necessary to distinguish the dimly lit limb from the background sky. Video cameras with only moderate light sensitivity may require recourse to a smaller image scale.

Recording Time-Lapse Sequences on Videotape

If you are making a time-lapse recording on videotape and do not intend to capture selected frames using a computer for further processing, centering the subject on the X becomes especially critical. Without the post-capture processing capabilities provided by a computer's software, you won't be able to adjust or crop the individual frames.

To ensure smooth, seamless transitions between each recorded sequence, use the Pause command to stop and start while in the record mode. Using the Stop command may result in unsightly gaps between sequences. In many VCRs the Pause command leaves the tape in a virtual paused mode for about five minutes, the maximum period of time that the videotape can rest against the heads before running the risk of damage. This feature facilitates almost immediate reactivation when you resume recording. But if five minutes elapse without any activity, the VCR will automatically release the tape from its paused or virtual stop and move it away from the record/playback heads. If you were to then press Record, in all probability the tape would not resume recording in synchronization with the last recorded frame due to imperfect mechanical tolerances and slight tape stretching. The result will be an objectionable noisy gap in the recording.

Be sure to activate each successive Pause and Record command at precisely measured intervals of time (for example, pause for 90 seconds, record for five seconds, pause for 90 seconds, repeating the sequence). Once the event is over, you can review your completed time-lapse recording in either the normal Playback mode or in the Fast Playback mode for the most dramatic effect.

Saving Image Sequences to a Computer

Whether you transfer images from videotape to your

computer using a frame grabber or you capture them "live" directly to your computer, you must designate each image file with a logical name and number. Each AVI movie sequence should be given an appropriate file name (e.g., jup001.avi, jup002.avi, jup003.avi, etc.) so that you will be able to sift through each AVI file in the proper sequence to select the sharpest frames with the best seeing.

Each single frame selected from an AVI movie sequence (or simple series of snapshots) should also be sequentially numbered and saved in accordance with its particular source movie (e.g., jup001.bmp, jup002.bmp, jup003.bmp, etc.). This will enable your computer's movie-editing software to assemble the frames in the proper sequence.

To avoid the loss of your valuable raw image files as well as those that you may have devoted hours to processing, create three separate folders. Name the first folder "Raw" and use it to store the raw image files; name the second folder "Processed" and use it to store the processed (enhanced) image files; and name the third folder "Movie" and use it to store the image files from the "Processed" directory that have been cropped and registered, ready for assembly into a complete movie that will be written to disk.

This discipline will allow you to refer to the original image or movie file in the event that you make a mistake. If you fail to do this, you'll have to obtain new originals from the initial AVI sequence. If you've already discarded that in order to save disk space, you're in deep trouble and will have to start over!

Processing the Images

If you modify the properties (brightness, contrast, hue, etc.) of the first image that will appear in your movie, make sure that you apply the same modifications to all the other images in the sequence. In some instances, a particular image may be a little brighter or darker than the previous ones due to the camera's automatic gain control, inconsistent shutter speed setting, or changing atmospheric conditions (seeing, transparency). If this occurs, open a previously processed image from the sequence in a new window to serve as a benchmark.

Place subsequent images beside it to make a careful visual or histogram comparison, then adjust them to match the benchmark image as closely as possible.

Some image-processing programs like Adobe Photoshop have macro recorders that remember the various enhancement routines applied to a reference image. This combination of routines can be assigned a special function key (such as F3) so that all the other images in the sequence can be identically modified with a single keystroke, saving vast amounts of time. Save your processed images in the Processed directory.

Image Registration and Cropping

Now it's time to precisely adjust the registration of each frame. At the telescope you kept your subject centered directly beneath the X on the screen of your monitor. However, buffeting of the telescope by wind, tracking errors caused by periodic error in the drive gears, or misalignment on the celestial pole may cause the features in your final images to appear in slightly different locations from frame to frame, especially when the frames are minutely examined at the pixel level. These subtle displacements can cause features to jump around or oscillate when the frames are assembled and played back as a movie. To avoid this annoying effect, each frame in a movie must have the same overall height and width, with stationary features (like lunar craters) in precise registration from one frame to the next. Fortunately, nearly all image-processing software programs are capable of displaying the horizontal and vertical (X and Y) pixel coordinates of features as you move the mouse over an image (Figure 16.4).

Let's take the case of a time-lapse movie of the shrinking shadow of Mount Piton as the Sun rises over this prominent lunar peak (Figure 16.4). After determining the desired overall pixel dimensions for the first frame of the movie, crop it accordingly and save it as a processed/cropped file in your movie folder. At this point you can name it frame001.tif, since the numbered sequence will automatically correspond to the source files in the Raw and Processed folders.

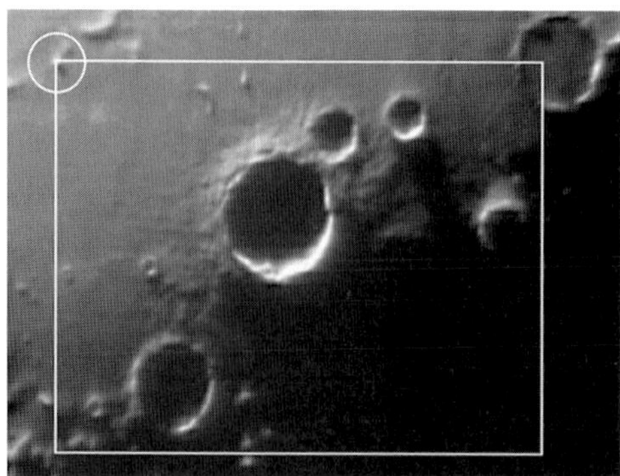

FIGURE 16.4
A lunar mountain (circled at upper left) *selected to serve as a reference point for cropping the frames of a time-lapse animation.*

Now you need to prepare the second frame in the sequence. With the first image still open, move it to one side of the screen and open the second image in the sequence. In this example it will be piton002.tif from the Processed folder.

A sophisticated image editor like Adobe Photoshop or PaintShop Pro will allow you to copy the second image and paste it over the first image (frame000.tif) as a new layer. Using the Transparency or Opacity options, you should set the value for the new layer (Layer 1) to 50 percent. The background image will be partially visible through it (as in Figure 16.6 with the planet Jupiter). Using the Move tool or the arrow keys, superimpose the fixed features in the image with the corresponding features in the background image. Visually, the image will take on a sharp appearance when matching features are properly aligned. If you are still uncertain, alternately press keys 1 and 9 to rapidly shift the opacity levels between the background image and Layer 1. Any obvious motion will confirm a slight misalignment, with the sole exception of the shadow, which will be somewhat shorter in the second image.

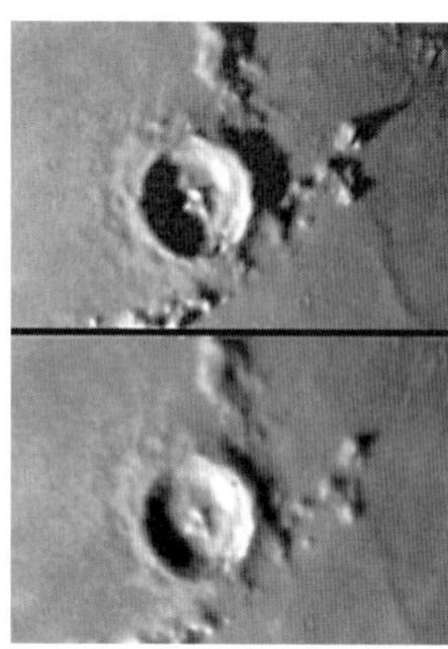

FIGURE 16.5
The crater Eratosthenes is another excellent subject for a time-lapse movie. A nearby mountain peak under midmorning lighting produces a shadow shaped like a witch's hat.

Once the frames are in precise registration, change the opacity of Layer 1 to 100 percent (or zero transparency) and effectively make the background image appear nonexistent. Now use the Flatten command to produce a new single layered image. The new image will now assume the same overall pixel dimensions as the former background reference image. This feature handily eliminates the need to fuss with the time-consuming cropping routines of many earlier image editors. Save it in the Movie folder as frame002. Repeat these steps for all the other frames in the movie sequence.

In the case of the planets and whole-disk recordings of the Sun (to show rotation or the stages of a solar eclipse) and Moon (to show libration or the monthly cycle of phases), the fixed reference point for aligning the images will be the edges of their disks. A feature on the lunar limb will serve when making a movie of an occultation. In all of these cases, the steps for achieving proper image registration described in the example of the Mount Piton movie still apply.

When you superimpose and align the transparent layers for the frames of a movie showing a planet's rotation, bear in mind that the planet's markings will appear a little out of focus even when the edges of the disks in the layers are in precise registration (Figure 16.6). Don't be alarmed — this unavoidable effect is caused by the displacement of the rotating planet's markings between the successive images, and this is exactly what we want. The same holds true for shifts in the position of the Galilean moons caused by their orbital motion.

FIGURE 16.6
Two 50 percent transparent images of Jupiter are aligned with image-processing software to create a well-centered time-lapse animation.

Completing the Movie

Before you assemble your movie, make a final check to ensure that all the frames are in the proper sequence and that their overall appearance after image processing is consistent throughout the series. Your movie-editing software should have a function called Convert Image Sequence, which automatically collates images in their assigned numerical order.

Adobe Premier and Ulead Video Studio are both excellent programs for editing and creating movies. With Ulead's Video Editor function, after you click on the first image in the sequence, the program tallies the total number of consistent frames. This number should equal the number of frames that you have created. If it doesn't, there is a problem, often an incorrectly named file or a file in a different format or of a different size than the others. Should this occur, make the necessary changes to pull the vagabond image back into the ranks. In the Options menu, you can select the movie's compression method and frame playback rate. Experiment with different playback rates until you find one that displays the event most effectively.

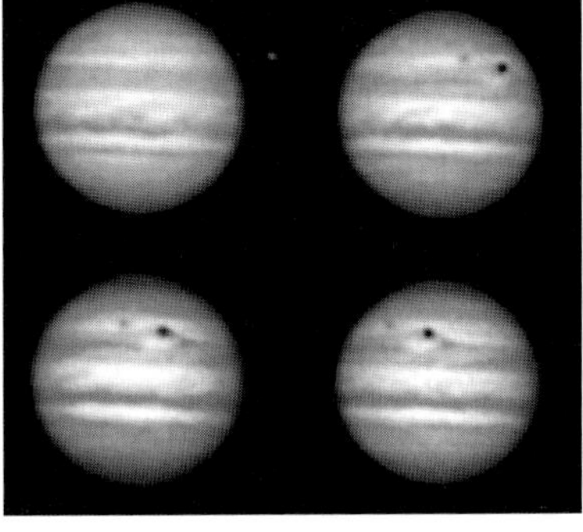

FIGURE 16.7
These images of a transit of Ganymede were captured on September 22, 1998, using eyepiece projection on a 10-inch f/4.5 Newtonian reflector. The approximate times (Universal Time) for the frames in this sequence are 10:15; 11:15; 11:58; and 13:10. The time-lapse animation was created using 180 frames. In the second frame (top right), Ganymede's tiny, dusky disk can be seen just to the left of the Great Red Spot, while the satellite's trailing shadow rests on its upper right edge.

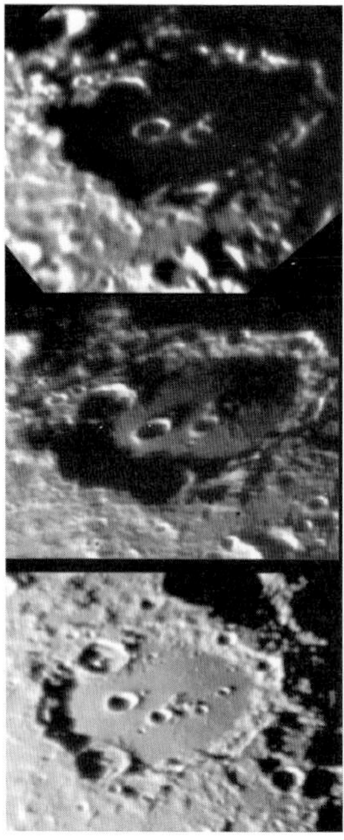

FIGURE 16.8
Three views of the imposing crater Clavius a few days after First Quarter. Note the arc of secondary craters that pop into view as the Sun rises over its huge convex floor.

The time-lapse movie can now be played back at any time using your preferred media player software. It will make a memorable presentation at a meeting of your local astronomy club. For those who want to display movies on a Web page, a variety of programs specifically tailored to create animated GIF files are available. With improvements in compression algorithms over the years, the animated GIF file has made its mark on the Internet world, particularly in the form of advertisement banners. But GIF-specific animations don't always produce the smooth, seamless-looking movies possible in AVI, particularly with larger images. Another format referred to as MPEG (the type used in today's DVD technology) offers varying levels of image preservation in a highly compressed format. MPEG produces movies of the same duration and of virtually the same quality as AVI, but at a fraction of the file size. This can be a real boon to visitors to your Web page, enabling them to download a copy in a matter of minutes for viewing (Figures 16.7, 16.8).

Below is a list of lunar features with sufficient topographic relief to make excellent subjects for time-lapse movies of retreating shadows at sunrise or advancing shadows at sunset.

FEATURE	LONGITUDE	LATITUDE
Apennine Mountains	West 3°	North 20°
Aristillus	East 1°	North 34°
Arzachel	West 2°	South 18°
Copernicus	West 20°	North 10°
Eratosthenes	West 11°	North 14°
Moretus	West 5°	South 70°
Mount Pico	West 9°	North 46°
Mount Piton	West 1°	North 41°
Petavius	East 60°	South 23°
Piccolomini	East 32°	South 29°
Plato	West 9°	North 52°
Teneriffe Mountains	West 13°	North 48°
Theophilus	East 26°	South 12°
Tycho	West 11°	South 43°

Appendixes

The Sequential Movie Player

This program is written for *Visual Basic 4* and is also compatible for versions up to *Visual Basic 6.0*.

Application Design

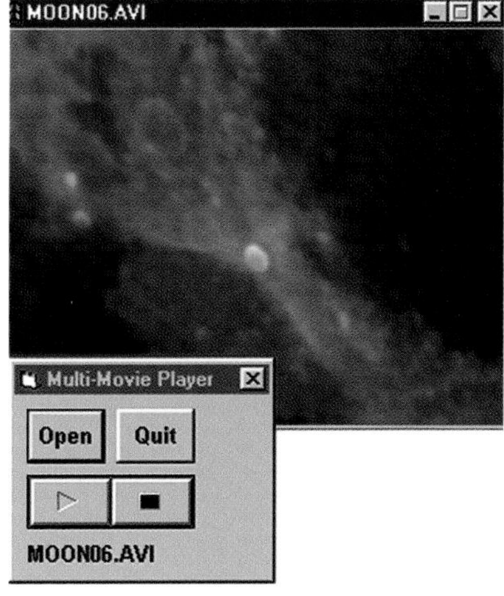

FIGURE A1.1
The Multi-Movie program allows any number of AVI files to be played back in sequence, eliminating the need to laboriously open each image file one at a time.

1. Open your Visual Basic program with a new blank form and project window.

2. Go to the toolbar and select the Tools option and Custom Controls. The Custom Controls window will pop up.

3. In this window check the available controls for the Microsoft Multimedia Control. If it is shown but not selected, select it now. If the control is not listed then click the Browse button and go into your Windows system directory. You will see a file called MCI32.OCX. Select this file and click on the Open button. This will now be added to your list of Custom Controls. Click OK. Your Toolbox window will now also display an additional icon for the media control tool. If you're using *Visual Basic 6.0*, right-click on the Tools window and select the Components option. Scroll down the list and select Microsoft Media Control and Microsoft Common Dialog.

4. On your blank project form, draw the following five controls. They are 2 command buttons side by side, one common dialog to the right, one Media Control below these, and a Label below the Media Control.

5. Select the first Command1 button, right click on it, and select Properties. Change its caption name to Open. The same procedure applies to Command2, though its caption will be Quit.

6. Apply the same process to Label1, only this time delete its default caption.

7. Right click on the Media Control button and

enter the Properties option. You will notice two selectable option columns on the right. In the leftmost column deselect everything other than the Auto Enable option, which should be ticked. On the right side column, tick the Play Visible and Stop Visible options and deselect the others. Click OK and your Media Control buttons will show only the Play and Stop buttons.

8. You can now size the form as desired placing all the controls where you please. The following code can be reproduced in Notepad. Then using the Select All and Copy functions, paste the code into the VB project form Declarations under General. The code will then automatically be assigned to the relevant functions.

9. Test the project by pressing F5 or click the Run button on the VB toolbar. Click on the Open button and select the first movie in your movie directory. Click OK and let the show begin! The movies should automatically play the full sequence of each AVI file, then unload it and load the next file. To end watching the current sequence of each movie, press the "Stop" button on the media control. This will unload the current movie and immediately load the next in the sequence.

10. If all appears to work correctly, you can create a stand alone executable file by selecting the File - Make EXE file options from the VB toolbar.

Multi-Movie Player Code

```
Private Sub Command1_Click()
CommonDialog1.Filter = "AVI|*.avi"
CommonDialog1.ShowOpen
ChDir CurDir
Label1.Caption = CurDir
FileSpec$ = "*.avi" 'Get filename
Label1.Caption = Dir$(FileSpec$)
On Error Resume Next 'Setup error handling.
MMControl1.Visible = True
MMControl1.Enabled = True
MMControl1.Notify = False
MMControl1.Wait = True
```

```
MMControl1.Shareable = False
MMControl1.DeviceType = "AVIvideo"
MMControl1.filename = Label1.Caption
'Open the MCI AVI video device
MMControl1.Command = "Open"
MMControl1.Command = "Play"
End Sub
Private Sub Command2_Click()
End
End Sub
Private Sub MMControl1_Done(Notify_Code As Integer)
MMControl1.Command = "Close"
On Error resume Next ' Setup error handling
Label1.Caption = Dir$
MMControl1.filename = Label1.Caption
'Open the MCI AVI video device
MMControl1.Command = "Open"
MMControl1.Command = "Play"
End Sub
```

Assembling a Compact Video Camera

II

The heart of any CCD video camera is the tiny circuit board containing the CCD array sensor and its supporting circuitry. If you're a do-it-yourself type, it's worth noting that these circuits, often designated as "board cameras," can be purchased from the suppliers listed in Appendix IV. A vast array of models are available in both black-and-white and color, offering moderate to high resolution and lux ratings as low as 0.02 lux. They are usually supplied with removable lenses. You may be surprised at the tiny dimensions of many of these circuit boards. It's possible to construct a compact video camera that weighs less than many telescope eyepieces!

FIGURE A2.1

A simple-to-build, light-sensitive astronomical video camera, based on an inexpensive surveillance camera circuit board. Housed in a lightweight enclosure and fitted with a machined adapter to fit the focuser of a telescope, cameras like this one are capable of producing astounding images of the Moon and planets.

The savings that result from assembling a video camera using one of these circuit boards and a readily available metal or plastic enclosure rather than purchasing a complete camera are marginal at best. However, an old camcorder with a broken tape-transport mechanism may be prohibitively expensive to repair, but more often than not its CCD circuit will still be functional and can be salvaged.

Board camera modules fit easily into the confines of small electronics project boxes. When selecting an enclosure, make sure that there is adequate clearance for the power input and video output connectors. In a couple of hours, using a soldering iron and a few simple tools, you can assemble a camera specifically designed for use with your telescope.

Some suppliers even offer handy die-cast metal or injection-molded plastic enclosures complete with predrilled holes. You can simply secure an adapter tube to the enclosure using a fast-setting epoxy cement. Connections are made using a small in-line four-way connector and preterminated cable instead of conventional panel-mount connectors.

FIGURE A2.2

Smaller than a book of matches, this low-light board camera contains all of the electronics required for producing full motion video.

Components

A basic component list would include:

1. An RCA or BNC panel-mount socket for

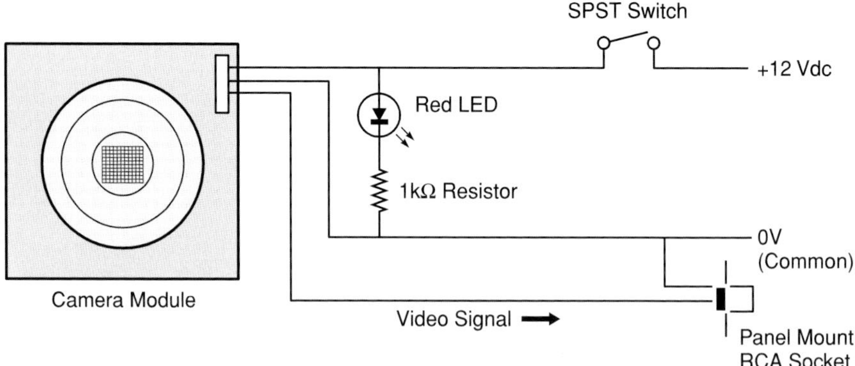

FIGURE A2.3

The wiring diagram for connecting the board camera and additional components.

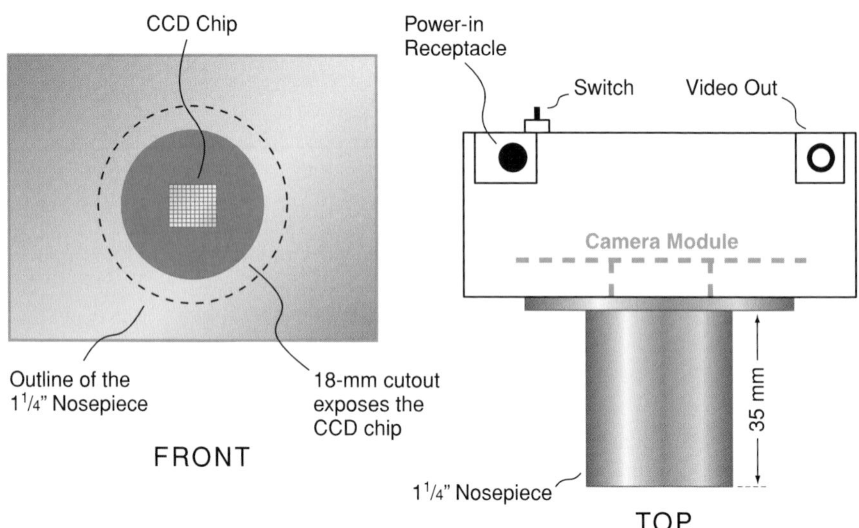

FIGURE A2.4

The cutouts required to permit light to reach the CCD array sensor and to mount additional components like connectors and switches.

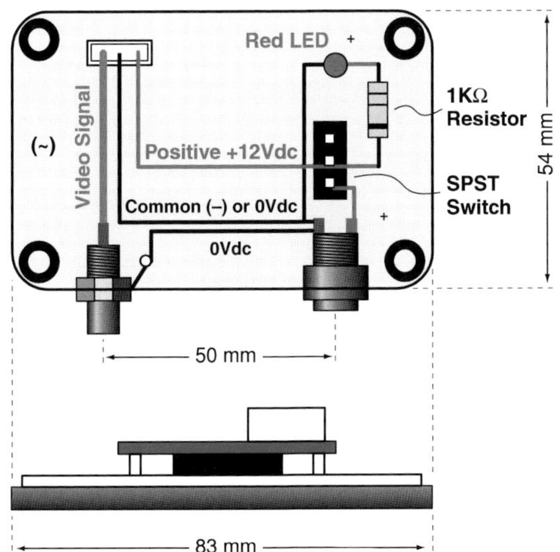

FIGURE A2.5
Internal wiring configuration for assembling the project box. Care must be taken to avoid short circuits.

composite video output.

2. A DC power input receptacle.

3. A 12V/150mA DC-regulated plug pack adapter.

4. A short length of 1.25- or 2-inch diameter aluminum or PVC tubing cut or machined to around 40 mm in length. Like the tube of an eyepiece, this will serve as the coupling for the camera and telescope focuser.

5. Optionally, a miniature SPST on-off toggle switch, with a 1K (1,000 ohm) resistor and red light-emitting diode (LED) to serve as a power indicator light.

Preparing the Video Board

If the video module you are using is just a bare circuit board, it will probably have predrilled holes at the corners so that it can be mounted using standoffs. Be careful not to touch any of the circuit board components, holding it only by the edges. This will reduce the possibility of static electricity permanently damaging the circuit's sensitive semiconductors. Static charges are especially problematic when the humidity is low and can have fatal effects on electronic components.

Wiring Up the Circuit

Video camera circuits are generally supplied with a small wiring assembly or loom. This will have a 3- or 4-

pin polarized connector at one end and prestripped, tinned wires at the other. If this has not been supplied, follow any enclosed wiring instructions carefully. These three wires are for the power input and video signal output. The red wire will connect to 12-volt positive, while the black wire will connect to the negative or ground terminal of the power supply or battery.

The wire that will carry the video output signal will most likely be a little thicker than the others, with a central wire surrounded by shield of copper foil or braid. This shield isolates the inner signal-carrying wire from stray electrical noise. It is connected to signal ground and will most likely be hard-wired together with the black (negative) wire at the connector on the circuit board or in the connector itself. If this is not the case, a common ground can be achieved using the battery or the negative terminal of the power supply at the power input connector. Keep a reasonable distance between the power and video output connectors to minimize noise from the power supply that would interfere with the video output signal.

Be sure to check the polarity of all wiring to the circuit board. Incorrect wiring of the (+) positive and (–) negative leads may result in permanent damage to the video circuit. If wiring instructions are provided with the module, be sure to read them carefully.

The optional LED must also be wired to the correct polarity. The positive or +12-volt wire connected via the switch's on position applies a voltage to the current-limiting resistor, which is connected to the positive or anode lead of the LED. This lead can be easily identified since it is slightly longer that the cathode lead, which connects to negative or ground.

Assembling the Camera Housing

Most video board cameras are supplied with a small lens in a threaded mount. These optics will not be required after the assembled camera is tested and can be removed at that time. It is best to leave it in place for the time being in order to protect the CCD array from dust and moisture.

An opening must be made in the project box enclosure's lid or cover plate to provide a path for light to reach the CCD chip on the video board. Using a hole saw will avoid splintering a plastic enclosure.

The adapter tube that will be used to couple the camera to your telescope's focuser can be machined from a short length of rigid PVC pipe obtained at the local hardware store. On a lathe, turn the outside diameter a few thousandths of an inch less than 1.250 inches. Alternatively, the adapter can be machined from aluminum or brass and with a mounting flange that will permit it to be attached to the enclosure using machine screws, facilitating any minor centering or perpendicularity adjustments that may prove necessary. In either case, the adapter needs to fit smoothly but securely in the barrel of your telescope's focuser.

Paint the interior of the tube flat black to minimize stray light. To further reduce glancing-angle reflections, consider cutting a fine thread (32 or 40 threads per inch) on the interior while the tube is chucked in the lathe.

If you've decided to construct a camera by purchasing a predrilled enclosure, the task of adapting it to your telescope's focuser is simply a matter of purchasing a C-mount adapter, which can be attached to the enclosure using epoxy cement. This step should only be performed after carefully insuring that the adapter is precisely centered with respect to the CCD chip on the circuit board.

Testing the Camera

Before you seal the enclosure, make sure that there are no bare wires or crossed connections that might result in damage to the video circuit components. Before connecting the video card to the circuit, use a voltmeter to check for correct wiring polarity and voltage levels.

Now secure the circuit board to the housing and seal the assembly. With the video output connected to your monitor or TV, apply power to the camera and watch the screen. You should see an image, though odds are it will be out of focus. Point the camera at various light sources around the room. The image displayed on the monitor should respond accordingly. If so, the camera is functioning correctly.

If your board camera came equipped with a lens, check to make sure that the camera's adapter tube is centered with respect to the CCD chip by focusing on an object one or two meters away. The vignetted image displayed on the monitor should be evenly illuminated and in uniform focus. If this is not the case, adjust the centering or perpendicularity of the adaptor tube accordingly.

Camera Storage and Maintenance

Always handle your video camera with great care and store it safely in a padded camera road case or similar padded enclosure.

The principal enemy of your CCD chip is dust. If even a tiny mote of dust settles on the surface of the array, it will look like a black boulder on the monitor. The best way to deal with dust is to prevent it from ever reaching your equipment. When not in use, keep your camera in a sealed container or store it face down on a scrupulously clean surface. An even better option is to purchase a Wratten #1A skylight filter that can be taped or cemented into your lathe-turned adapter (or simply be screwed into most commercial C-to-1.25-inch adapters). This is an extremely effective way to exclude dust.

Should dust get onto your chip's surface, you'll need to clean it. You will need a small camel's hair brush, some detergent and distilled water, and a can of compressed air from the local camera store. The first step in getting rid of dust is to attempt to blow it away with gentle jets of compressed air. This has little chance of scratching the delicate surface of the CCD array. If this fails, take the camel's hair brush and dip it repeatedly in a glass of tap water containing a couple of drops of dishwashing soap, kneading the hairs a few times to ensure that they are free of oil or grime that may have been picked up from someone's hands. Rinse the brush thoroughly with tap water, followed by distilled water. Let the brush dry thoroughly, then gently fluff its bristles and lightly brush the surface of the chip.

Resources and Suggestions for Further Reading

III

1: Clearer Skies Through Video Eyes

Dantowitz, Ron. "Sharper Images Through Video," *Sky & Telescope*, Vol. 96 (August 1998), 48–54.

Gross, Todd. "You, Your Weather, and Your Skies," *Sky & Telescope*, Vol. 99 (January 2000), 125–30.

MacFarlane, Alan W. "A Primer for Video Astronomy," *Sky & Telescope*, Vol. 79 (February 1990), 226–31.

MacRobert, Alan. "Beating the Seeing," *Sky & Telescope*, Vol. 89 (April 1995), 40–43.

2: How Video Cameras Work

Luther, Arch C. and Andrew F. Inglis. *Video Engineering*, 3rd ed. New York: McGraw-Hill, 1999.

3: Selecting a Video Camera

The VideoAstro e-mail discussion group's Web site contains numerous postings that compare the performance of various video cameras, as well as galleries of amateur video images: **www.fortunecity.com/victorian/canterbury/222/astrovid.htm**

Buchanan, John. "QuickCam Astronomy," *Sky & Telescope*, Vol. 95 (June 1998), 120–23.

Dyer, Alan. "SBIG's STV Integrating Video Camera and Autoguider," *Sky & Telescope*, Vol. 101 (January 2001), 67–73.

Horne, Johnny. "The Astrovid Color PlanetCam," *Sky & Telescope*, Vol. 102 (August 2001), 55–59.

_____. "Four Low-Cost Astronomical Video Cameras," *Sky & Telescope*, Vol. 105 (February 2003), 57–62.

Moore, David M. "Video of the Stars." *Sky & Telescope*, Vol. 98 (August 1999), 61–64.

4: Video Signal Processors

Luther, Arch C. and Andrew F. Inglis. *Video Engineering*, 3rd ed. New York: McGraw-Hill, 1999.

Inoue, Shinya. *Video Microscopy*. New York: Plenum Press, 1986.

5: Choosing a VCR

Luther, Arch C. and Andrew F. Inglis. *Video Engineering,* 3rd ed. New York: McGraw-Hill, 1999.

McComb, G. and A. Rathbone. *VCRs and Camcorders for Dummies,* 2nd ed. New York: IDG Books, 1995.

6: Video Monitors

Luther, Arch C. and Andrew F. Inglis. *Video Engineering,* 3rd ed. New York: McGraw-Hill, 1999.

7: Videocassettes, Cables & Connectors

Luther, Arch C. and Andrew F. Inglis. *Video Engineering,* 3rd ed. New York: McGraw-Hill, 1999.

McComb, G. and A. Rathbone. *VCRs and Camcorders for Dummies,* 2nd ed. New York: IDG Books, 1995.

8: Combining Video with Computers

Berrevoets, Cor. "Processing Webcam Images with *RegiStax,*" *Sky & Telescope,* Vol. 107 (April 2004), 130–35.

Kelly, Al, Richard Berry, Ed Grafton, and Chuck Shaw. "True-Color CCD Imaging," *Sky & Telescope,* Vol. 96 (December 1998), 142–47.

Ratledge, David, ed. *The Art and Science of CCD Astronomy.* London: Springer-Verlag, 1999.

Aguirre, Edwin L. "Shooting the Space Station with Video," *Sky & Telescope,* Vol. 104 (December 2002), 132–38.

9: Hard Copy

Dobbins, Thomas A. "Shoot the Moon and Planets!" *Sky & Telescope,* Vol. 91 (June 1996), 94–97.

Inoue, Shinya. *Video Microscopy.* New York: Plenum Press, 1986.

Harper, Rick. "Hard Copy." *CCD Astronomy,* Vol. 1 (Spring 1994), 22–25.

10: The Moon

The Lunar Section of the Association of Lunar and Planetary Observers (ALPO) welcomes contributions from amateur astronomers: **www.lpl.arizona.edu/alpo**

Cherrington Jr., Ernest H. *Exploring the Moon Through Binoculars and Small Telescopes.* New York: Dover, 1984.

Dobbins, Thomas A. "Recording the Moon and Planets with a Video Camera," *Journal of the British Astronomical Association,* Vol. 106 (December 1996), 309–14.

_______. "A Lunar Video Gallery," *Journal of the Association of Lunar and Planetary Observers,* Vol. 41 (1999), 137–39.

Massey, Steve. "Luna-Cam," *Sky & Space,* Vol. 10 (June/July 1997), 55–57.

Price, Fred W. *The Moon Observer's Handbook.* Cambridge: Cambridge University Press, 1988.

Rükl, Antonín. *Atlas of the Moon,* revised, updated ed. Cambridge, Massachusetts: Sky Publishing Corporation, 2004.

Wood, Charles A. *The Modern Moon: A Personal View.* Cambridge, Massachusetts: Sky Publishing Corporation, 2003.

11: The Planets

The Association of Lunar and Planetary Observers (ALPO) offers valuable resources for planetary observers, including a training program: **www.lpl.arizona.edu/alpo**

Beatty, J. Kelly, Carolyn Collins Petersen, and Andrew Chaikin, eds. *The New Solar System,* 4th ed. Cambridge, Massachusetts: Sky Publishing Corporation and Cambridge University Press, 1999.

Davis, Michael and David Staup. "Shooting the Planets with Webcams," *Sky & Telescope,* Vol. 105 (June 2003), 117–22.

Dobbins, Thomas A., Donald C. Parker, and Charles F. Capen. *Observing and Photographing the Solar System.* Richmond, Virginia: Willmann-Bell, 1988.

Doherty, Paul. *Atlas of the Planets.* New York: McGraw-Hill, 1980.

Legault, Thierry. "Thoughts on High-Resolution Imaging," *Sky & Telescope,* Vol. 99 (January 2000), 148–52.

Massey, Steve. "Leader of the Opposition," *Sky &*

Space, Vol. 12 (August/September 1999), 52–54.

Numazawa, Shigemi. "Using a CCD on the Planets," *Sky & Telescope*, Vol. 83 (February 1992), 209–15.

Price, Fred W. *The Planet Observer's Handbook,* 2nd ed. Cambridge: Cambridge University Press, 2000.

Robinson, Leif J. "Saturn and a Winking Star," *Sky & Telescope*, Vol. 78 (September 1989), 259.

Schaaf, Fred, *Seeing the Solar System.* New York: John Wiley & Sons, 1991.

Troiani, Daniel M. and Daniel P. Joyce. "A Camcorder Assist for Planetary Observers," *Sky & Telescope*, Vol. 80 (October 1990), 409–10.

12: Occultations & Transits

Detailed instructions on video equipment and techniques for recording occultations are posted on the Web site of the International Occultation Timing Association (IOTA): **www.occultations.org**

The Occultations page on *Sky & Telescope*'s Web site provides up-to-date information on lunar, asteroidal, and planetary occultations as well as camcorder timing tips: **SkyandTelescope.com/observing/objects/occultations**

Aguirre, Edwin L. "Photographing the Transit of Venus," *Sky & Telescope*, Vol. 107 (May 2004), 137–41.

Maunder, Michael and Patrick Moore. *Transit: When Planets Cross the Sun.* London: Springer-Verlag, 2000.

Povenmire, Harold. *Graze Observer's Handbook,* 2nd ed. Indian Harbour Beach, Florida: JSB Enterprises, 1979.

13: Eclipses

Aguirre, Edwin L. "Imaging Totality," *Sky & Telescope*, Vol. 98 (July 1999), 136–41.

Espenak, Fred. *Fifty Year Canon of Lunar Eclipses: 1986–2035* (NASA Reference Publication 1216). Cambridge, Massachusetts: Sky Publishing Corporation, 1989.

_______. *Fifty Year Canon of Solar Eclipses: 1986–2035* (NASA Reference Publication 1178 Revised). Cambridge, Massachusetts: Sky Publishing Corporation, 1994.

Harrington, Philip S. *Eclipse! The What, Where, When,*

Why and How Guide to Watching Solar & Lunar Eclipses. New York: John Wiley & Sons, 1997.

Littmann, Mark, Ken Wilcox, and Fred Espenak. *Totality: Eclipses of the Sun,* 2nd ed. New York: Oxford University Press, 1999.

Pasachoff, Jay M. and Fred Espenak. "Videotaping the Eclipse," *Sky & Telescope*, Vol. 82 (July 1991), 103–4.

14: The Deep Sky

The Web site of the QuickCam and Unconventional Imaging Astronomy Group (QCUIAG) offers lots of information on long-exposure imaging with modified security video cameras, webcams, and many more: **www.qcuiag.co.uk**

Anton, Rainer. "Measuring Double Stars with Video," *Sky & Telescope*, Vol. 104 (July 2002), 117–20.

Ashford, Adrian R. "Deep-Sky Imaging with Integrating Video Cameras," *Sky & Telescope*, Vol. 106 (December 2003), 131–34.

Chambers, Stephen and Stephen J. Wainwright. "Deep-Sky Imaging with Webcams," *Sky & Telescope*, Vol. 107 (January 2004), 137–42.

Di Cicco, Dennis. "Intensifying Your Viewing Experience," *Sky & Telescope*, Vol. 97 (February 1999), 63–66.

Horne, Johnny. "StellaCam II: Taking Video into the Deep Sky," *Sky & Telescope*, Vol. 108 (October 2004), 86–89.

Saulietis, Andy and Paul Maley. "A Compact Image Intensifier," *Sky & Telescope*, Vol. 76 (December 1988), 632–33.

15: Meteors

An extensive posting of articles on observing meteors using video can be found at the International Meteor Organization's Web site: **www.imo.net**

Bone, Neil. *Meteors.* Cambridge, Massachusetts: Sky Publishing Corporation, 1993.

Bone, Neil. *Observing Meteors, Comets, Supernovae, and Other Transient Phenomena.* London: Springer-Verlag, 1999.

Hawkes, R. L. "Constructing a Video-Based Meteor

Observatory," *Journal of the International Meteor Organization,* 18–4 (1990): 145.

Molau, Sirko. "MOVIE – Meteor Observation with Video Equipment," *Proceedings of the International Meteor Conference 1993,* (1994): 71.

_____, M. Nitschke, M. de Lignie, R. L. Hawkes, and J. Rendtel. "Video Observations of Meteors: History, Current Status, and Future Prospects," *Journal of the International Meteor Organization, 25–1* (1997): 15.

_____. "Automated Meteor Observing," *Sky & Telescope*, Vol. 101 (May 2001), 132–36.

Okamura, Osamu. "Aerial Rendezvous with the Leonids," *Sky & Telescope*, Vol. 101 (May 2001), 137.

Povenmire, Harold R. *Fireballs, Meteors, and Meteorites.* Indian Harbour Beach, Florida: JSB Enterprises, 1980.

Ueda, M. and Y. Fujiwara. "Television Meteor Radiant Mapping," *Earth, Moon, and Planets, Vol. 68* (1995): 585.

Equipment and Supplies IV

This section presents a list of companies that sell the products discussed in this book. Product categories are arranged alphabetically and numbered, and the relevant numbers appear below the vendors' names and addresses.

1. Astronomical Image-Processing Software
2. Astronomical Video Cameras
3. Board Cameras and Enclosures
4. Camcorder-to-Telescope Adapters
5. Desktop Computer Video Cameras (Webcams)
6. Digital Video Interface and Firewire Products
7. Filters and Filter Wheels
8. Flip Mirrors
9. Frame Grabbers and Video Capture Boards
10. High-Resolution-Format VCRs
11. Image Intensifiers
12. Security and Surveillance Video Cameras
13. Video Monitors
14. Video Printers
15. Video Signal Processors

Adirondack Video Astronomy
72 Harrison Ave.
Hudson Falls, NY 12839
www.astrovid.com
1, 2, 8, 9, 11, 13

All Things Sales & Services
P.O. Box 110
Kelmscott 6991
Western Australia
www.allthings.com.au
3, 12, 13

Apogee Inc.
P.O. Box 136
Union, IL 60180
www.apogeeinc.com
8

Audio Video Supply
4575 Ruffner St.
San Diego, CA 92111
www.avsupply.com
3, 9, 10, 12, 13, 14, 15

Celestron
2835 Columbia St.
Torrance, CA 90503
www.celestron.com
7

Cohu, Inc.
Electronics Division
P.O. Box 85623
San Diego, CA 92186
www.cohu-cameras.com
12

Collins Electro Optics LLC
9025 East Kenyon Ave.
Denver, CO 80237
www.ceoptics.com
11

Colorado Video, Inc.
2100 Central Ave., Suite 109
Boulder, CO 80301
www.colorado-video.com
15

DV411
3767 Overland Ave., Suite 103
Los Angeles, CA 90034
www.dv411.com
6

Diffraction Limited
25 Conover St.
Ottawa, ON K2G 4C3
Canada
www.cyanogen.com
1

Edmund Industrial Optics
101 East Gloucester Pike
Barrington, NJ 08007
www.edmundoptics.com
9, 11, 12, 13, 14, 15

Herbach and Rademan
353 Crider Ave.
Moorestown, NJ 08057
www.herbach.com
3, 11, 12, 13

ITE Astronomy
16222 133rd Dr. N
Jupiter, FL 33478
www.iteastronomy.com
2

JAI PULNiX, Inc.
1330 Orleans Dr.
Sunnyvale, CA 94089
www.pulnix.com
12

Lechner Electric CCTV
Pirschweg 16
D-83071 Stephanskirchen
Germany
www.lechner-cctv.de
2

Logitech, Inc.
6505 Kaiser Dr.
Fremont, CA 94555
www.logitech.com
5

Lumicon International
750 Easy St.
Simi Valley, CA 93065
www.lumicon.com
7

Meade Instruments Corporation
6001 Oak Canyon
Irvine, CA 92618
www.meade.com
7, 8

Murnaghan Instruments Corporation
1781 Primrose Lane
West Palm Beach, FL 33414
www.e-scopes.cc/Murnaghan
_Instruments_Corp56469.html
7, 8

Optec, Inc.
199 Smith St.
Lowell, MI 49331
www.optecinc.com
7, 8

PC Connection
Route 101A, 730 Milford Rd.
Merrimack, NH 03054
http://shop.pcconnection.com
5, 9

Perseu, Unipessoal Lda.
Rua Dr. Agostinho Neto, 1 1° D
2695-395 Sta. Iria da Azóia
Portugal
www.perseu.pt
2

Phil Dyer
www.mintron.co.uk
2

PocketScope. com LLC
4945 Evergreen Valley Way
Alpharetta, GA 30022
www.pocketscope.com
5

Polaris Industries, Inc.
3158 Process Dr.
Novcross, GA 30071
www.polarisusa.com
2

Santa Barbara Instrument Group
147-A Castilian Dr.
Santa Barbara, CA 93117
www.sbig.com
1, 2

ScopeTronix
1423 SE 10th St., Unit 1A
Cape Coral, FL 33990
www.scopetronix.com
4

Sky Publishing Corp.
49 Bay State Rd.
Cambridge, MA 02138
SkyandTelescope.com
1

Supercircuits, Inc.
One Supercircuits Plaza
Liberty Hill, TX 78642
www.supercircuits.com
3, 12

Taurus Technologies
P.O. Box 14
Woodstown, NJ 08098
www.taurus-tech.com
8

Telescopes & Astronomy
P.O. Box 292
Ohalloran Hill, SA 5158
Australia
www.telescopes-astronomy.com.au
5

True Technology Ltd.
Woodpecker Cottage, Red Lane,
Aldermaston, Berks RG7 4PA
England
www.trutek-uk.com
7, 8

Van Slyke Engineering
12815 Porcupine Lane
Colorado Springs, CO 80908
www.observatory.org
7, 8

Video Direct Electronics
5450 NW 33rd Ave., Suite 110
Fort Lauderdale, FL 33309
www.video-direct.com
10

Vixen Co. Ltd.
5-17 Higashitokorozawa
Tokorozawa, Saitama 359-0021
Japan
www.vixen.co.jp
2

Watec Co. Ltd.
c/o GENWAC, Inc.
60 Dutch Hill Rd., Suite 6
Orangeburg, NY 10962
www.genwac.com
www.watec.net
12

Willmann-Bell, Inc.
P.O. Box 35025
Richmond, VA 23235
www.willbell.com
1

Image Credits

V

1.1,1.2, 1.3, 1.6, 1.7, 1.9: Steve Massey; 1.5: Gregg Dinderman; 1.8: Dennis di Cicco; 1.10, 1.11, 1.12: Ron Dantowitz

2.1, 2.6, 2.7: Steve Massey; 2.2, 2.3, 2.4: Gregg Dinderman and Lick Observatory; 2.5: Murnaghan Instruments

3.1, 3.4: Steve Massey; 3.2: Craig M. Utter; 3.3: Adirondack Video Astronomy; 3.5: Tasco (Australia); 3.6: Philips; 3.7: Tan Wei Leong; 3.8, 3.9, 3:10: Steve Wainwright; 3.11: Santa Barbara Instrument Group

4.1: Eric Douglass

4.2: Steve Massey

5.1: Steve Massey; 5.2: Ron Dantowitz

6.1: Dennis di Cicco; 6.2, 6.3, 6.4: Gregg Dinderman

7.1: Craig M. Utter; 7.2: Gregg Dinderman

8.1: Belkin; 8.2, 8.3, 8.4, 8.6, 8.7, 8.8, 8.9, 8.10, 8.11, 8.12, 8.13, 8.15, 8.16: Steve Massey; 8.5, 8.14: Gregg Dinderman; 8.17: Murnaghan Instruments

9.1, 9.4: Steve Massey; 9.2, 9.3: Craig Utter

10.1, 10.2, 10.3, 10.5, 10.6, 10.7, 10.8, 10.9, 10.16, 10.17, 10.19: Steve Massey; 10.4, 10.10, 10.11, 10.12, 10.13, 10.14, 10.20, 10.21, 10.22, 10.24: Thomas A. Dobbins; 10.15, 10.18: Charles Genovese Jr.; 10.23: Consolidated Lunar Atlas; 10.25: Audoin Dollfuss

11.1: Gerald Stelmak; 11.4A, 11.4B, 11.5, 11.6, 11.7, 11.8, 11.11, 11.14: Steve Massey; 11.2, 11.3, 11.10, 11.12, 11.13: Ron Dantowitz; 11.9: David Moore; 11.15: Leif Robinson and Dennis di Cicco; 11.16, 11.17: Steve Massey

12.1: Dale Ireland; 12.2, 12.3, 12.4, 12.7: Ron Dantowitz; 12.5: Lorenzo Comolli; 12.6: Steve Massey

13.1, 13.5: Steve Massey; 13.2: Imelda B. Joson; 13.3: Gordon Garcia; 13.4: Ron Dantowitz

14.1, 14.2, 14.3, 14.5, 14.6: Steve Massey; 14.4: Dennis di Cicco

15.1: Gregg Dinderman; 15.2: Steve Quirk and Rob McNaught; 15.3: Sirko Molau; 15.4, 15.5: David Palmer

16.1–16.8: Steve Massey

A1.1: Steve Massey

A2.1: Steve Massey; A2.2: Oatley Electronics Australia; A2.3, A2.4, A2.5: Gregg Dinderman

Index (Page numbers in boldface type indicate figures.)